FIELD MANUAL
No. 34-8-2

Headquarters
Department of the Army
Washington, DC, 1 MAY 1998

Table of Contents

DISTRIBUTION RESTRICTION: Approved for public release; distribution is unlimited.

LIST OF TABLES AND FIGURES

PREFACE

This publication is a roles and missions manual for G2/S2 sections. As the companion handbook to FM 34-8, this manual is written for officers serving as the G2 or S2 in combat, CS, and CSS units. It provides guidance for officers and NCOs assigned to G2/S2 sections. G2s should find this handbook a useful MTP when executing intelligence training responsibilities.

The proponent of this publication is the United States Army Intelligence Center and Fort Huachuca, Fort Huachuca, AZ. Send comments and recommendations on DA Form 2028 (Recommended Changes to Publications and Blank Forms) directly to Commander, US Army Intelligence Center and Fort Huachuca, ATTN: ATZS-TDL-D, Fort Huachuca, AZ 85613-6000.

This manual does not implement any ISAs. However, it complies with the following STANAGs:

- STANAG 2022, Intelligence Reports, Edition 8.
- STANAG 2077, Order of Battle, Edition 5.
- STANAG 2149, Intelligence Requests, Edition 5.
- STANAG 2844, Counterintelligence Procedures, Edition 2.
- STANAG 2936, Intelligence Doctrine, Edition 2.
- STANAG 6010, Electronic Warfare in the Land Battle, Edition 1.

It also complies with QSTAG 593, Mutual Support Between EW Units; and QSTAG 1034, Intelligence Preparation of the Battlefield.

Unless specifically designated, all references to the intelligence officer at corps, division, brigade, or battalion will be stated as S2.

Acronyms and brevity codes used in this manual are identified only in the glossary.

Unless this publication states otherwise, masculine nouns and pronouns do not refer exclusively to men.

Chapter 1

G2/S2 ROLES AND MISSIONS

WHY YOU NEED THIS HANDBOOK

The G2/S2 must provide continuous intelligence and information for the commander to conduct operations and minimize risk.

Intelligence is the commander's decisionmaking tool. The commander, who drives intelligence, does so for planning before deployment, while enroute, and during operations and redeployment. The S2/NCO—

- Provides the commander timely and accurate intelligence, IPB, I&W, and vulnerability assessments for force protection, targeting, and BDA.
- Makes decisive predictions on when and where an action will take place.
- Prioritizes IR.
- Integrates with other staff elements on I&S issues.
- Provides the commander with a view of all facets of the battlefield.
- Ensures his staff is trained.

PURPOSE OF THIS HANDBOOK

This manual does not replace the doctrine and TTP contained in the other FM 34-series manuals; it does, however, focus on their application. It also summarizes information that helps the S2 manage and coordinate the CCIR. It provides the S2 the roles and missions required for executing the intelligence support function.

The G2/S2 is the "Top Down"/"Bottom Up" integrator of RISTA operations, which have always been performed at the brigade and battalion levels. Three factors help to redefine traditional R&S planning:

- Top Down reporting.
- Bottom Up reporting.
- Digitization.

Systems and technologies available to units make an integrated approach to managing "top down" and "bottom up" reporting feasible. The ACT serves as the nexus of RISTA operations, digitally linking the brigade S2 with the automated capabilities of the DS MI company, resulting in RISTA operations which blend intelligence, R&S, and TA. (See FM 101-5 for G2/S2 responsibilities and Table 1-1 for RISTA collection resources.)

Table 1-1. Collection resources

IEW	RISTA
SIGINT	Troops in contact
IMINT	Other troops in contact (RAS)
HUMINT	Scouts (aerial & ground)
CI	LRS, SOF
MASINT	Countermortar & CB radars
TECHINT	Gun cameras
	Hand emplaced sensors
	CA, PAO, PSYOP

Chapter 2

FORCE PROJECTION

FORCE PROJECTION PRINCIPLES

Future Army operations will rely more heavily on the force projection of US combat power. There is no single method to support force projection. This chapter identifies key principles and considerations for planning and executing IEW force projection.

Successful force projection of IEW support is based on understanding and applying the key principles shown in Figure 2-1.

Other key force projection components are intelligence readiness and a requirement to define responsibilities each echelon performs down to the lowest tactical level. This chapter discusses—

- Intelligence readiness.
- I&W.
- Stages of force projection operations (minus operations).

Considerations are in a logical order but may not be sequential nor applicable for all force projection missions. Corps and division G2s must provide primary planning support for force projection operations because of their expertise. The S2 at each level identifies—

- Personal knowledge deficiencies (e.g., communications architecture, imagery dissemination systems and methods, ability to identify the staff's shortcomings).

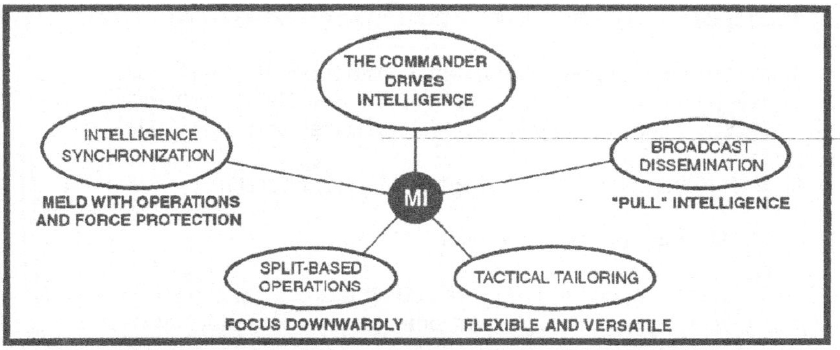

Figure 2-1. Principles of force projection IEW operations.

- Technical limitations (e.g., ASAS, SIDS capability).
- Intelligence gaps with adequate specifics.

INTELLIGENCE READINESS

In a JTF force projection operation, higher echelons will provide intelligence for situation and target development to lower echelons (top down) until the tactical ground force completes entry and secures the lodgment area. The JTF J2 may be reluctant to push everything down through tactical intelligence channels due to the volume of the intelligence information available. The S2 may receive support on a "smart push" basis, and needs to know his requirements to be able to do a "smart pull."

The most significant change in the evolution of force projection operations is the enhanced information flow through hierarchical and non-hierarchical networks (computer, communications, and personal). The S2 should—

- Review available databases on assigned contingency AIs; conduct IPB on these AIs; and develop appropriate IPB products.
- Be aware of higher HQ SOPs and DIA manuals for specific CM guidance.
- Prepare, practice, and conduct information collection activities as part of in-garrison IO rehearsals.

- Preplan and practice an intelligence "surge" on likely contingency crises.

- Prepare and practice coordination from predeployment through redeployment with personnel from imagery, SIGINT, HUMINT, SWO, CA, PSYOP, and SF units, to include databases and connectivity.

- Ensure the following are a part of the daily operating environment:

  — RC and other augmentation.

  — Line numbers and SOPs including a linguist plan with proficiency scores (alert through early entry phases of deployment).

  — Training (individual and collective).

- Form ad hoc intelligence links and networks early on to meet a developing contingency. Incorporate, request, and receive intelligence from unfamiliar sources (linguists, MI augmentation, other services); exploit NGOs and PVOs once a crisis emerges. Exchange communications protocols with theater and higher HQ and subordinate and lateral units.

- Forward all RIIs to the higher HQ IAW SOPs. The J2, G2, and S2 can focus intelligence downward based on the commander's needs.

- Understand the JTF J2's multiple echelon and broadcast dissemination capability to ensure NRT reporting to all deployed, in transit, or preparing to deploy forces.

- Maintain an intelligence database on the battlefield environment and threats for each contingency. The S2 must state and record the CCIR as PIR, subordinate SOR, and SIR, and include the following:

  — For the battlefield environment, the commander's approval of the AI, to include separate ground, air, littoral waters, and political AIs.

  — Maps, terrain, and weather products. Request from National Imagery and Mapping Agency hard copies (unclassified or at the lowest classification). Request authority to declassify these products locally.

— Digitized products (map sheets for ASAS, terrain data, and imagery).

— Physical environmental information. The TERRA BASE program allows S2s to template the effects of terrain on communications and direct fire. During mission analysis, TERRA BASE or other automated terrain products—WINCATS, TOPOSKINNER—provide the S2 a tool to help the commander visualize how terrain can affect friendly and enemy forces. These products can illuminate terrain effects for subordinate commanders in the OPORD brief. [Technique: Use the program at Home Station to develop and sustain proficiency.] The supporting engineer staff officer may also have terrain visualization products.

— Threat or potential threats. The intelligence community, primarily the NGIC, and open sources produce products useful for intelligence readiness. These products can be tailored to best support the commander. (See FM 34-3 for information on intelligence analysis.) INTELLINK will be an S2's primary access to any type of requested strategic intelligence. Examples of strategic level products include—

  - Global security forecast.
  - Battlefield development plans.
  - Automated and hardcopy databases.
  - Arms proliferation and military power studies related to the weapons acquisition strategies and the overall military power and potential of selected foreign military forces.
  - TECHINT and User Bulletins.
  - CIA World Fact Book and DIA country studies.
  - Open source studies and articles.
  - Other services.

INDICATIONS AND WARNINGS

Theater and national intelligence units monitor regional and global threats to provide I&W intelligence to the NCA and military commanders. I&W intelligence flows to strategic, operational, and tactical commanders; it prevents surprise, reduces risk, and supports development and refinement of CONPLANs. The S2 must ensure the commander identifies PIR, IR, and targeting requirements for each assigned contingency area. The S2 should—

- Conduct CM and synchronization planning on I&W requirements. Review your unit's collection plan and preplanned SOR for each contingency area. If necessary, refine existing collection plans and SOR.
- Review and modify reporting procedures for I&W contingency areas. This may involve changing intelligence reporting (e.g., increasing reporting on one area and decreasing reporting on another) and message routing addresses and precedence (e.g., FLASH designation).
- Prioritize and forward SOR to higher headquarters.
- Disseminate intelligence and information to the commander, staff, and subordinate units.
- Coordinate for direct dissemination when possible.
- Recommend to the commander whether to maintain or increase unit readiness levels; plan and surge the intelligence effort for the impending operation; or move the unit from its current mission to contingency, branch, or subsequent operations.
- Adjust intelligence readiness steps according to pre-crisis I&W.

STAGES OF FORCE PROJECTION OPERATIONS

REMINDER: All phases of the intelligence cycle are being executed continuously during all stages of force projection.

MOBILIZATION:

Mobilization is a process in which the armed forces augment the AC capability in preparation for war or other national emergencies. It includes activating all or part of the RC assembling and organizing personnel, supplies, and materiel; and certifying the proficiency of individuals and units.

(See FM 100-17 for mobilization process.) In peacetime, AC and RC MI units plan, train, and prepare together to accomplish mobilization and deployment tasks. During the mobilization phase, the S2 should—

- Assist the mobilizing RC unit by preparing and conducting intelligence training and threat update briefings and by disseminating intelligence.
- Identify force requirements for the different types of operations and CONPLANs.
- Employ and adhere to existing procedures.
- If possible, use prior coordinated IMAs to fill gaps created by personnel shortages. These IMAs should already have a working knowledge of your SOPs and understand the mission.
- Monitor intelligence reporting on threat activity and I&W indicators.
- Conduct or coordinate CI and OPSEC training and operations.
- Manage IR and RII from your unit and subordinate units.
- Evaluate reporting.
- Update collection planning.

PREDEPLOYMENT ACTIVITY:

Predeployment activity provides the foundation for subsequent force projection operations.

- I&W will continue throughout force projection operations. The S2 must perform IPB early and continuously. (See FM 33-1, FM 34-130, FM 41-10.)
- Accelerated training will ensure all AC and RC MI units are able to conduct IEW operations. Integrate mobilization and deployment tasks into unit METL and training; emphasize and integrate critical aspects of force projection into battle tasks and planning.
- Focus broadbased knowledge to support CONPLAN refinement (IPB).

- Ensure tactical tailoring or split-based operation planning is based on an existing CONPLAN.

- Establish appropriate relationships; establish higher, lower, and lateral liaisons if they are different.

- Continuously conduct and update CM and IEW synchronization planning.

  — IO synchronization ensures IO are linked to the CCIR and respond in time to influence decisions and operations. IO encompasses more than just MI assets and requires MI support. (See FM 34-40 and FM 100-6.)

— IEW synchronization ensures IEW operations are linked to the CCIR and respond in time to influence decisions and operations. IEW is a subset of IO.

— The commander generates CCIR. CCIR focus on what is critical. They should change as the situation changes. CCIR affect PIR, IR, EEFI, and FFIR.

o Plan imagery coverage for target nomination, validation, and PSA.

o Require from collection assets, timelines for preplanned and dynamic collection requests.

o Plan synchronization through all five steps of the intelligence cycle.

o Ensure each PIR is related to a specific operational decision. (See FM 34-2 and FM 100-6.)

o Prepare an ISM and backward plan so collection production efforts are executed with the operation; deliver focused intelligence to support operational decisions.

o Coordinate with the S3 and signal officer for EW offensive and electronic deception operations, specifically for target nomination, no-fire target criteria, protected frequencies, and synchronized EW effort during all phases of the operation.

o Plan BDA requirements.

o Refine, manage, and update SOR. Monitor and maintain synchronization.

- o Anticipate and initiate collection early against long lead-time requirements.

- o Ensure CCIR process is continuous and the intelligence cycle and IEW operations remain tied to the commander's decisions and concept of operations.

- Identify collection gaps after you synthesize available information on the AO and coordinate the collection efforts of existing intelligence organizations. (Consider national, EAC, other US forces and services, and HN support.)

- Develop a collection strategy which factors PIR, IR, and METT-TC. Submit RII and other requests for support to adjacent, higher, and HN units. Develop a collection plan which supports all PIR and IR and maintains synchronization.

- Develop your intelligence team:

— *Commander's intelligence support* The S2 and supporting MI commander form the maneuver commander's intelligence team. ASAS is the primary intelligence processing system supporting this team. A unit may need to augment its S2 staff, to include area experts, TECHINT LNOs, linguists, interrogators, intelligence analysts, and CI agents. Tailor the size and sophistication of the deploying unit's intelligence staff to the mission.

— *ACT*: The DS MI company provides an ACT to support the commander's intelligence team at brigade.

  — *Additional attachments* These could include PSYOP and CA.

  — *DISE*: This is a small intelligence support team that provides communications, automated intelligence fusion, and broadcast downlinks in a small package capable of deploying with Army early entry forces.

    - o Decide whether you will form the DISE from the organic assets of the early entry force or from the corps ACE, supporting EAC MI brigade, or a combination thereof.

    - o Plan the configuration, which can range from "briefcases" (Mini-DISE) to HMMWVs (DISE vehicular), and are normally staffed by 5 to 12 soldiers, respectively. Build the supporting hardware systems around the Army's ASAS and integrate them with other Army, joint intelligence, and communication capabilities.

- o Plan whether the DISE will expand to a full ACE as lodgment operations are completed or will disband once the mission is accomplished.

- o Plan situations in which the DISE will go directly to higher headquarters for information without the home base's approval.

- o Plan on requirements to support the DISE with 24-hour operations.

— *ISB*: This is the rear element of split-based operations that provides processed and analyzed intelligence to the DISE.

- o Plan the automation, communications capacity, and personnel necessary to provide continuous intelligence requirements management, collection, processing, and reporting.

- o Plan all procedures for the DISE to "pull" specific intelligence products and reports, obtain status of collection, and "push" the current intelligence picture from the lodgment back to the ARFOR and JTF commanders if not in theater.

— Plan all links to higher intelligence organizations, including the JIC and the elements designed to leverage theater intelligence; focus on support down to the corps and echelons below corps (e.g., the CMISE).

— Coordinate with and understand the capabilities of the JTF to include the NIST if formed.

— Plan to use RC augmentation. Some examples are a mix of CI agents, UAVs, interrogators, and linguists.

- o Transportation availability (aircraft or naval vessel) for deployment.

- o Sustainability.

- o Portability once deployed.

- o Disciplined operations. The commander's ability to collect may be affected by nonmilitary decisionmakers. The CINC can contribute to the deploying commander by ensuring at the outset that intelligence is decompartmented and releasable to allied units (if applicable).

— Review applicable publications (ARs, DA Pamphlets, USSIDs).

— Review SOFAs, ROE, international laws, and other agreements. (Coordinate with SJA on these issues.)

— Review applicable NIMA guidance.

— Establish force deployment priorities based upon METT-T. Sequence initial required forces and capabilities, build-up priorities, and follow-on forces to ensure a sequenced plan, a tailored force, and established command and support relationships. Consider sensors, processors, preprocessors, CI, and HUMINT. Maintain unit integrity.

— Plan communications architecture (build redundancy when possible). Remember to—

  - Plan links to the JTF DISE, if formed, that complements the NIST. (See FM 34-25-3 and FM 34-37).

  - Ensure intelligence links provide the early entry commander vital access to multisource Army and joint intelligence collection assets, processing systems, and databases.

— Ensure collection is synchronized with production, and intelligence is still synchronized with operations (specify reporting procedures and timelines).

— Finalize the IEW OPLAN (terrain and communications deconfliction). Coordinate with the MI commander on his tactical decisionmaking process. Understand the MI commander's SOR that are his specified tasks, implied tasks, task organization, concept of operation (the organization, deployment, allocation, and employment of subordinate MI units), and coordination requirements with forward maneuver units.

— Establish the intelligence crossover point. Estimate the time and establish a measurable criteria to indicate when you have reached that point. Intelligence crossover occurs when enough tactical

collection assets are in theater to reduce the dependency upon strategic or national assets. (See FM 34-1, Chapter 3.)

- Receive all augmentation and support elements quickly (ACT, DISE) and incorporate them within your unit, SOP, and training.
- Update databases to support the IPB process that will follow. Practice using INTELLINK.
- Support force protection. Intelligence operations—MDCI in particular—identify, locate, and target an enemy's ability to target and affect friendly forces, facilities, and operations. Intelligence support must—
  - — Conduct threat and risk assessment.
  - — Consider elements of fratricide avoidance:
    - o Accurate target identification.
    - o Collateral weapon effects.
    - o Familiarity with supporting units.
    - o ROE.
  - — Consider elements of force protection:
    - o Information Warfare:
      - * OPSEC.
      - * Physical destruction.
      - * PSYOP.
      - * Deception.
      - * EW.
    - o COMSEC.
    - o NBC.

- Personnel security.
- Neutrality maintenance.
- Personal awareness.
- Sniper threat.
- Counterreconnaissance.
- Arms and physical security.

— Using the MDCI process, assess and review friendly vulnerabilities and the threat's ability to exploit them.

- S2s should inform commanders and operators on CI and MDCI analysis.
- Train your MDCI analysts to conduct reverse IPB and think like the enemy S2, to include the enemy's perception of friendly centers of gravity and how he will attack or influence them.
- Incorporate your MDCI analysis into the G2/J2 situation and decision briefings and all planning (especially deception planning).
- Assign CI and MDCI sections appropriate missions and analytical responsibilities, such as rear area threat analysis, and continuously assess effectiveness.

— Identify potential countermeasures to deny the enemy access to friendly critical areas.

— Identify and recommend actions to counter enemy intelligence collection capabilities.

— Implement the following controls to support force projection:

- Establish access to national HUMINT and CI databases, automated links to joint service, coalition, and HN sources to help identify, assess, and develop countermeasures for threats. Develop FLASH precedence reporting procedures.

  - Receive and disseminate CI information and specific CI tasks from ASAS and other means such as the CHATS.

- Support further CONPLAN and OPLAN development. MI units continually monitor and update their CONPLANs to reflect the evolving situation, especially during crises.

- Immediately before deployment, update deploying forces with the most recent intelligence on the AO and update your technical databases and situation graphics.

- Develop contingency tailored packages that allow the G2/S2 to place the right force support teams in a deployable posture with an adequate amount of training.

DEPLOYMENT:

Success in force projection operations hinges on the capability of airlift and sealift assets to move forces into the AO. Force protection is more critical during this stage.

- Monitor intelligence reporting on threat activity and I&W indicators. Continuously conduct and update IEW, CM, and synchronization planning.

- Plan enroute updates to eliminate information voids and allow your commander to adjust OPLANs prior to arrival in theater. Request supporting intelligence organizations use SATCOM, broadcast technology, and ADP systems to provide graphic and textual intelligence updates while enroute.

- Continue IPB.

- Provide timely, accurate, and specific infrastructure and weather information. IMETS can provide weather information.

- Use accurate situation development to help the commander understand and reduce risk on the battlefield.

- Use the DST, collection plan, R&S plan, ISM, and SOR to anticipate which decisions the commander and staff will make.

ENTRY OPERATIONS:

During initial entry operations, EAC organizations provide major intelligence support in a "smart push" mode. This support may include departmental, joint, and scalable Army intelligence elements capable of deploying forward. Entry units must continue to "smart pull" the intelligence they need for operations.

- Continue to conduct force protection planning.

- Monitor the buildup of the in-theater capability required to conduct sustained IEW operations and to reduce your total dependence on split-based, "top driven" intelligence from outside the AO. As organic IEW assets flow into the theater, assess their reliability for tactical intelligence. (National and theater organizations will still remain sources of strategic intelligence.) Determine intelligence crossover point.

- Monitor intelligence reporting on threat activity and I&W indicators.

  — Routinely debrief troops.

  — Use local nationals employed by or in frequent contact with your force to provide information (area experts must analyze it).

- Ensure liaison personnel and basic communications are in place prior to the scheduled arrival of parent commands.

  — Deploy HUMINT forces into theater as early as possible.

  — Contact HN or civil authority. (Contact CA units; see FM 41-10.)

  — Contact other services.

  — Contact lateral units.

  — Coordinate with supporting CI unit for CI support to force protection.

- Emplace ACT, DISE, NIST, and other elements.

  — Establish security.

  — Establish communications.

- — Establish analytical capability.
- — Perform IPB.
- — Continue to conduct CM and intelligence synchronization planning. Recommend revised PIR and IR. Refine, manage, and update SOR. Evaluate reporting.
- — Consider space requirements, power, and logistical support for high use or unique items.

- Conduct situation development, target development, and support to targeting. During this stage as combat strength increases, your unit's organic tactical systems will conduct situation and target development (intelligence crossover point).

- Develop measurable criteria to evaluate the results of your collection plan. Reassesses—
  - — "Push" versus "pull" requirements.
  - — Communications architecture.
  - — Reporting procedures and timelines.
  - — Crossover point in intelligence.
  - — Intelligence support to OPLANs and OPORDs, branches, and sequels (to include planning follow-on forces).

WAR TERMINATION AND POSTCONFLICT OPERATIONS:

Upon cessation of hostilities or truce, deployed forces transition to a period of postconflict operations.

Commanders redirect PIR and IR to support units conducting restoration operations (e.g., engineer units conducting infrastructure reconstruction operations, medical and logistics units providing humanitarian relief). The nature of the PIR shifts from assessing threat forces to assessing political, economic, social, religious, and other conditions that affect force protection and the desired end-state; planning residual presence of US forces; and preparing for redeployment of forces.

- Continue to conduct force protection planning.
- Remain open to the possibility that hostilities could resume.
- Monitor intelligence reporting on threat activity and I&W indicators.
- Continue to conduct CM and intelligence synchronization planning. Update collection planning.
- Refine, manage, and update SOR.
- Evaluate reporting.
- Disseminate intelligence.
- Use pre-deployment tactical tailoring procedures to plan a phased redeployment of IEW assets and personnel ensuring continual coverage of the commander's requirements.

REDEPLOYMENT AND RECONSTITUTION:

As the combat power and resources decrease in the AO, force protection and I&W become the focus of the CCIR. This drives the selection of those MI units that must remain deployed and those which may redeploy.

- Monitor intelligence reporting on threat activity and I&W indicators.
- Continue to conduct force protection planning.
- Request intelligence BOS support (theater and national systems) and provide intelligence in support of redeployment and reconstitution (reverse intelligence crossover point).

DEMOBILIZATION:

Demobilization is the process by which MI individuals and units transfer from active to a premobilization or other approved posture. MI units resume intelligence readiness posture. RC MI units transition to peacetime status.

- Monitor intelligence reporting on threat activity and I&W indicators.
- Capture consolidated databases.

- Capture lessons learned via AARs (doctrine and TTP).
- Maintain intelligence readiness (e.g., training).
- Adjust MTOEs and evaluate the need for IMAs.

Chapter 3

MILITARY DECISION-MAKING PROCESS (MDMP)

MDMP in General

- FM 101-5 is the doctrinal source.
- The MDMP is a single, established, and proven analytical technique.
- The commander must follow the one-third/two-thirds planning rule.
- MDMP helps the commander and staff examine a battlefield situation and reach logical decisions.
- The commander can decide to use the complete or abbreviated version.
- Staffs should train on both the complete and abbreviated versions.
- Incomplete execution of the MDMP is a recurring deficiency at the CTCs.

Intelligence in the MDMP

- MDMP is based on continuous IPB, especially initial IPB during mission analysis.
- The commander drives intelligence; IPB is an integrated staff function driven by the commander.
- Train your section to conduct IPB so you can coordinate closely with other staff and BOS representatives.
- You must understand how all the BOS operate and how to integrate intelligence during planning, especially within the targeting process IAW FM 6-20-10.
- "Push" the staff to develop a robust and integrated R&S plan.

Staff Estimates

- FM 101-5 provides a generic staff estimate.
- The generic staff estimate is the base for the commander's operations, personnel, intelligence, logistics, CMO, communications, and special staff estimates.
- These staff estimates are designed to form, analyze, compare, and recommend friendly COAs.
- Staff estimates are continuous, must not be overly time consuming, and do not have to be a written product (time dependent).
- The wargame results and staff estimates help the staff compare COAs.

The Intelligence Estimate

- The purpose, scope, and content of the intelligence estimate changed with draft of FM 101-5.
- The old intelligence estimate was a text-based product derived from IPB that focused on ECOAs, capabilities, strengths, and vulnerabilities.
- This old intelligence estimate is functionally replaced by IPB products.
- A standard staff estimate prepared by the G2/S2 is the new intelligence estimate—it is radically different.
- The new intelligence estimate focuses on the G2/S2's ability to support friendly COAs and is used to compare and approve friendly COAs.

Step 1: Receipt of Mission	Intelligence in Step 1
■ The mission comes from higher headquarters or is derived from an ongoing mission. ■ On receipt of a new mission, the G3/S3 issues a WARNO to the staff. ■ The staff immediately prepares for mission analysis (SOP preparation). ■ Immediately the commander and staff do a quick initial assessment with emphasis on an initial allocation of available time. ■ The commander issues his initial guidance, and the G3/S3 issues a WARNO to subordinate units.	■ Collaborate with your higher G2/S2 before and during receipt of mission to facilitate the initial IPB. ■ As part of the initial assessment, look for gaps in your intelligence database and products. ■ Proactively request/prepare terrain products, weather, light, and climatology data; update the MCOO and doctrinal templates; maintain the enemy situation. ■ Focus on the initial time allocation (complete or full MDMP), R&S guidance, and any additional tasks or focus from the commander's initial guidance. ■ Establish/verify an IHL.
Step 2: Mission Analysis 1. Analyze the higher HQ order. (If confused by or you disagree with it, seek immediate clarification or resolution.) 2. Conduct initial IPB. 3. Determine specified, implied, and essential tasks. (It is important to understand specific requirements for each task.) 4. Determine the AI. 5. Review available assets. (The staff must identify additional resources needed to ensure the mission's success.) 6. Determine constraints (normally found in the scheme of maneuver, concept of the operation, and coordinating instructions). 7. Identify critical facts and assumptions. (List all appropriate assumptions from higher; state relevant conditions over which the commander has no control.) 8. Conduct risk assessment. 9. Determine initial CCIR. (Limit to 10 or less.)	**Intelligence in Step 2** ■ Intelligence and intelligence-related products: ⇒ AI. ⇒ Initial PIR (from the commander.) ⇒ Initial OPSEC vulnerabilities and EEFI.* ⇒ MCOO and terrain as described by OCOKA and its effects. ⇒ Assumptions (include enemy mission, objectives, threat BOS activities). ⇒ Other elements of the battlefield. ⇒ Situation templates (unrefined). ⇒ Event templates and matrices (unrefined). ⇒ Center/centers of gravity. ⇒ HVTs. ⇒ The IPB portion of the mission analysis brief and associated graphics (OB, weaknesses and peculiarities, activities and capabilities, and COAs). ⇒ Collection plan (initial, G2 only).

10. Determine the initial reconnaissance plan. (The resulting R&S annex sets reconnaissance in motion.) 11. Plan use of available time. (The commander and staff refine initial plan for use of available time.) 12. Write the restated mission. Who, what, when, where, and why. 13. Conduct mission analysis briefing. (Given to commander and staff. The briefing is critical to ensure a thorough understanding of planning.) 14. Approve the restated mission. 15. Develop the initial commander's intent. A clear, concise statement of what the force must do regarding the enemy, terrain, and desired end-state. 16. Issue the commander's guidance. This provides additional guidance to focus staff planning. 17. Issue a WARNO. 18. Review facts and assumptions. When facts or assumptions change, the commander and staff must assess their impact.	⇒ The R&S plan (initial plan that starts R&S operations) and associated FRAGOs or WARNOs. ⇒ Intelligence estimate (initial). ■ These products are used to write Annex B to the OPORD or OPLAN and as the foundation for the DST* later in the MDMP. ■ Based on mission analysis, request information or intelligence based on intelligence gaps. ■ Develop as many ECOAs as time will allow. ■ Your higher HQ OPORD, OPLAN, and/or Annex B will task your unit to perform certain collection or R&S tasks. ■ Start the subsequent steps of CM & intelligence synchronization (FM 34-2) to support the initial collection plan (at a minimum SII, SIR, and SOR). ■ The R&S plan is a coordinated staff effort that must include FS, MEDEVAC, and CONPLANs (e.g., what the brigade does if 2 key scout sections are destroyed). ■ The MI Bn/DS MI Co participates in mission analysis and briefs the collection status and capabilities.
Step 3: COA Development 1. Analyze relative combat power. See FM 34-130 for estimating relative-force ratios. 2. Generate options. Goal is to develop COAs for every feasible ECOA; however, the commander usually limits that option with his guidance. 3. Array initial forces. Identify number of units needed and operational methods; develop a knowledge base to help make decisions. 4. Develop the scheme of maneuver. 5. Assign headquarters. This creates the task organization.	**Intelligence in Step 3** ■ Intelligence and intelligence-related products: ⇒ Situation templates (refined and prioritized). ⇒ Event templates and matrices (refined). ⇒ EWTL (initial).* ■ Ensure the G3/S3 uses the IPB facts, assumptions, and products developed during mission analysis and subsequently refined. ■ Work with the entire staff to ensure that friendly COAs take advantage of the environment and threat situation.*

6. Prepare COA statements and sketches. ■ The G3/S3 uses appropriate media to clearly portray how unit will accomplish the mission (e.g., scheme of maneuver).	■ Key on threat vulnerabilities. ■ Your input is critical in analyzing relative combat power. ■ Do not forget to support deception planning (when appropriate).
Step 4: COA Analysis (Wargme) 1. Gather the tools. 2. List all friendly forces. 3. List known assumptions. 4. List known critical events and DPs; include an HVTL list. 5. Determine evaluation criteria. 6. Select the wargame method. 7. Select a method to record and display results. 8. Wargame the battle and assess results. ■ The wargame is a critical and disciplined process used to visualize the flow of battle. ■ The commander selects the order of comparison of threat to friendly COAs. ■ The staff must evaluate the need for branches and sequels. ■ When technically possible, the staff should capture as much of the wargame on ATCCS as possible; otherwise, use a wargame worksheet or ISM. ■ These wargame results are key to developing DSTs and BOS synch matrix. ■ Use the action, reaction, counteraction method (consider at a minimum maneuver, FS, mobility, counter-mobility, survivability, and IEW). ■ The staff should track force ratios throughout the wargame.	**Intelligence in Step 4** ■ Intelligence and intelligence-related products: ⇒ PIR with LTIOV (refined). ⇒ HPTs. ⇒ Confirmation of the enemy center/ centers of gravity. ⇒ Situation templates (final). ⇒ Some force protection issues (e.g., NBC vulnerabilities). ⇒ EWTL (refined).* ■ You wear both a "red" and "blue" hat in wargaming. ■ As the enemy commander you project enemy actions or reactions, develop DPs, and project enemy losses. ■ During the wargame you must address all relevant enemy BOS capabilities. ■ As the friendly G2/S2, you identify IR and NAIs; refine the situation template; and participate in the targeting conference. ■ Ensure the G3/S3 honestly portrays friendly capabilities during the wargame.* ■ Work with entire staff to ensure friendly COAs take advantage of environment and threat situation. ■ Ensure HPTs, AGMs, and TSS support the operation.* ■ The AGM is approved by the commander and addresses which targets will be attacked, how, when, and the desired effects. ■ TSS are criteria used in deciding whether to pass information as a target nomination.

	■ The MI Bn/DS MI Co commander and collection manager are important players at the wargame.
Step 5: COA Comparison ■ Used to identify COA that has highest probability of success. ■ Staff may use any technique; the decision matrix is the most common. ■ Staff gets its criteria from the commander (e.g., the principles of war or tenets of Army operations).	**Intelligence in Step 5** ■ Intelligence product: ⇒ Intelligence estimate (final). ■ Ensure you play a key role during this step.
Step 6: COA Approval ■ If the commander modifies a proposed COA or gives the staff a new COA, staff must wargame that COA.	**Intelligence in Step 6** ■ Intelligence and intelligence-related products: ⇒ PIR with LTIOV (approved). ⇒ DST (an integrated staff product). ⇒ BOS synch matrix.* ⇒ ISM. ⇒ Event templates and matrices (final). ⇒ The R&S plan (final).* ⇒ Collection plan (refined, G2 only). ■ Start the subsequent steps of CM and intelligence synchronization (FM 34-2) (i.e., development of new SII, SIR, and SOR). ■ If the commander designates you to perform BDA to support one of his decisions, you must thoroughly plan the BDA support and tie that plan into the collection plan.
Step 7: Orders Production ■ Based on the commander's decision and final guidance.	**Intelligence in Step 7** ■ Intelligence and intelligence-related products: ⇒ OPORD or OPLAN Annexes B (Intelligence), L (R&S), P (C^2W)*, Q (OPSEC)*, S (Deception)*, and T (EW).*
* The G2 participates in developing the products but is not the proponent.	

ACCELERATING THE MDMP:

The MDMP involves three techniques that can be applied in different situations: deliberate, abbreviated, and accelerated. Only the accelerated technique is discussed herein because it is the most difficult to implement. (See CALL Newsletter Update No. 95-12.)

The accelerated technique may be used when one or more of the following conditions apply:

- Commander has a staff available to assist him in developing the plan, but little time to use a more formal process.

- Commander does not have a staff, or the staff is not accessible.

When these conditions apply, the commander must rely primarilyon TLP to develop his plan. The accelerated technique assists the commander in developing a tentative plan. Under extreme circumstances, this may be little more than a mental process; nonetheless, the commander can use it to assist him and key staff members (S2/S3/FSO/XO) as he develops his plan.

The accelerated technique follows the basic procedures in the deliberate and abbreviated processes, but the differences are more significant. The major differences between the abbreviated and accelerated techniques involve the commander's guidance and the COA development phase (Figure 3-1). The accelerated technique is characterized by active participation by the commander, and development of one COA that is suitable, feasible, and flexible.

In some situations, the products developed using the accelerated technique may be the same as those developed when using the deliberate or abbreviated technique. Because time is the key factor, the accelerated technique will normally result in the development of a FRAGO.

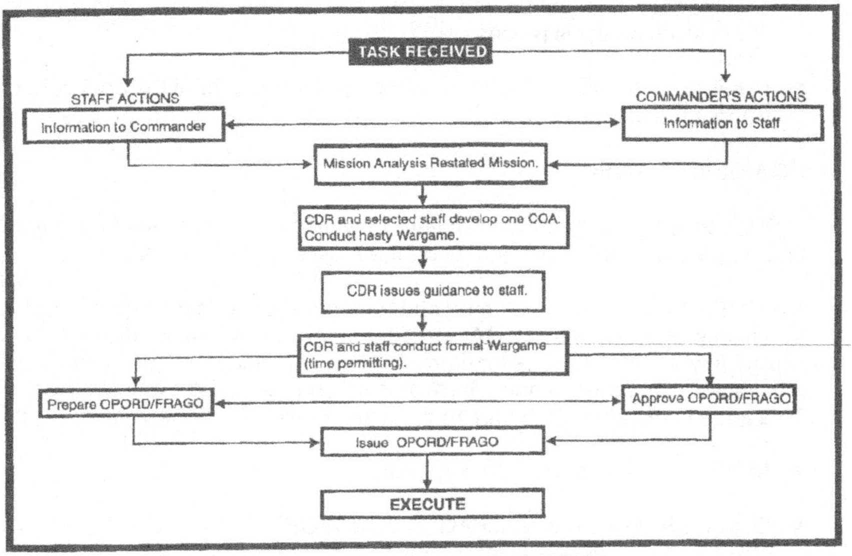

Figure 3-1. Accelerated MDMP.

MISSION ANALYSIS:

ISSUE: The commander and staff must be able to rapidly conduct the mission analysis to determine the restated mission.

DISCUSSION: When using the deliberate or abbreviated technique, the staff conducts a detailed mission analysis to develop the restated mission. As previously stated, resorting to a mental process in lieu of a detailed restated mission is acceptable; however, this should be the exception rather than the norm. The staff may be forced to brief their initial estimates orally, without the use of charts or viewgraphs. Conduct as formal a mission analysis as time allows. During the mission analysis, there are no major differences between the three techniques. There are no techniques that will significantly reduce the amount of time required to conduct the mission analysis. Anticipation, prior preparation, and experience by the staff are the keys to a timely mission analysis process.

TECHNIQUES:

- Commander must get personally involved by supervising and managing the mission analysis process.

- In extreme situations, the staff must be prepared to brief the commander without the use of visual aids.

COA DEVELOPMENT:

ISSUE: When time is severely limited, providing the commander's guidance after the mission analysis may not be the most appropriate time.

DISCUSSION: Instead, the commander may decide to immediately begin personally developing *one* COA with input from selected staff officers. There is probably not time to seek input from every staff officer, so the commander must determine relevant and critical staff officers (e.g., S2, F3, FSO, XO). This team may vary depending on the type of mission. For example:

- In the defense include staff engineer.

- During SASO include CA, PAO, SJA, PSYOP.

- In other situations, include subordinate commanders because of their experience.

This team must then quickly develop a flexible COA to accomplish the mission. The key to success, when using the accelerated technique, is to rapidly develop a base plan with appropriate branches that is flexible, feasible, suitable, and acceptable. *DO NOT WORRY ABOUT DEVELOPING THE PERFECT COA; THERE IS NOT TIME FOR IT* This is the major distinction between the accelerated technique and the others.

Once the COA is developed, the commander might consider conducting a hasty wargame. In extreme situations, this may be the only opportunity to conduct the wargame process. Next, the commander should begin to quickly develop his guidance to the staff. The accelerated technique is

characterized by an active role of the commander, and very specific guidance to the staff.

TECHNIQUES:

- Focus on developing one COA with branch plans that is flexible, feasible, suitable, and acceptable.

- The commander plays the central role when developing this COA.

COMMANDER'S GUIDANCE:

Once the commander has developed the COA, he must issue guidance to his staff so it can refine and wargame the COA. The commander's guidance to the staff must be directive and specific. The staff's responsibility is to support the commander's plan, not to develop the perfect plan. Well-developed and clearly communicated commander's guidance can be a significant timesaver. The commander's guidance should serve to keep the staff focused by establishing parameters to work within. Commander's guidance must be constantly reviewed and analyzed. As the situation changes and information becomes available, the commander may have to alter his guidance to the staff.

COA ANALYSIS:

ISSUE. The commander and staff must rapidly conduct the COA analysis process.

DISCUSSION: Conducting the wargame process using the accelerated technique is the most difficult of the three processes because only one COA was developed. The purpose of the COA analysis is not to analyze and compare multiple COAs that result in a recommendation to the commander, but to synchronize and integrate the commander's directed COA. This wargame session should focus on refining the branches or contingencies to the base plan and follow the formal wargame process as much as time allows. Focus on the most critical events. You do not have time to wargame the entire operation. When wargaming using the accelerated technique, the commander's involvement is even more important. The staff should use the box technique, focusing on actions at the objective or the engagement area. If time permits, wargame other critical events as well. The staff must work to support the commander's plan. However, as the staff works to refine the plan, it cannot become so biased that it develops a plan that is infeasible and insupportable. If the staff determines that it cannot support the commander's plan, then a new COA must be developed.

TECHNIQUES:

- If time permits, conduct a hasty wargame session during the COA development step. Ensure you identify and develop branches to the base plan.
- Involve the commander. He must supervise the wargame session, actively participate, make decisions, and provide guidance as required.
- Use the box technique, focusing on the most critical event first.

DECISION:

When using the accelerated technique, a decision brief is not required because only one COA was developed. The only decision that may be required is if the developed COA becomes unsuitable, infeasible, or unacceptable. If this occurs, another COA must be developed.

ADVANTAGES:

- Requires less time.
- Facilitates adaptation to a rapidly changing situation.
- Allows commander to compensate for lack of a staff or an experienced staff.

DISADVANTAGES:

- Significantly limits staff initiative and flexibility.
- Very directive, explores only one friendly COA.
- May result in only an oral order or FRAGO.

CHAPTER 4

S2 OPERATIONS CHECKLISTS

Intelligence operations will usually begin with a notice tasking statement. CONPLANs, training, the N-Hour sequence, and the MDMP contribute to mission readiness. The S2 can use the following checklists to monitor intelligence readiness prior to receipt of a mission and subsequently to verify preparations and facilitate mission planning.

The timelines used could apply to any echelon.

- Mission Checklist (Table 4-1).
- N-Hour Critical Times (Table 4-2).
- N-Hour Critical Actions List (Table 4-3).
- Mission Planning Sequence (Table 4-4).
- Reconnaissance Planning (Table 4-5).

Table 4-1. Mission checklist.

Mission Day Minus	ACTION	REMARKS
A-28	Notify attachments to provide updated access rosters prior to A-Day.	
	Provide open access to national and strategic databases.	
	Initiate verification of clearances within battalion.	Include sensitive positions and RTOs.
	Coordinate security briefings for unit personnel.	Include all OPSEC program components and SAEDA program.
	Coordinate contingency AI briefings.	
	Verify access to intelligence databases through division or higher.	Disseminate hard copy products to subordinate units.
	Inspect unit areas and equipment for physical security deficiencies.	Coordinate for support and access. Check equipment compatibility.
	Review section files.	Designate deployable and nondeployable records.
	Obtain current TECHINT/User Bulletins and DIA Top Ten Equipment Acquisition list from 203d MI Bn (TECHINT).	Acquire information on disposition directives from JCS and theater commander.
A-21	Obtain higher HQ access roster.	Update as required throughout mission cycle.
	Coordinate security force requirements with tasked units and MP.	
	Identify linguists in unit or update current file.	Provide list, crosscheck with S1 data.
	Provide updated list of telephone numbers to be cut off at N+2 to the S1.	
A-14	Finalize security plans and instructions.	Check guard/MP patrols.
A-8	Coordinate security requirements for filler personnel.	Barrier material, lighting, container seals, security of vehicle holding area, motor pool, empty barracks.

Table 4-1. Mission checklist (continued).

Mission Day Minus	ACTION	REMARKS
A-8	Prepare DA Form 3964 for classified material transported with deploying elements.	
	Brief MI battalion LNO on duties.	
A-7	Finalize MI battalion personnel and equipment support list.	Ensure copy placed in SDO book.
	Update section alert notification roster.	
	Continually monitor intelligence WATCHCON levels.	
A-4	Check alert rosters in SDO book and completion of all unit OPSEC and SAEDA briefings.	
A-Day	Brief RDC on battalion security plan.	EEFI material, handle accordingly.
	Issue guard instructions.	
	Provide access rosters to rear detachment personnel, higher and supported HQ.	

Table 4-2. N-Hour critical times.

N-Hour	EVENT	REMARKS
N-Hour	Notify and assemble units.	
	Initiate telephone control.	
	Initiate area security plan.	
	List questions for N+2 brief.	
N+1:30	Prepare staff equipment to go to staging area.	
	Verify basic load, load plans.	Maps, batteries, etc.
N+1:45	Depart for N+2 brief location.	
	Distribute updated access roster.	
	Secure battalion conference room.	
N+2:00	Present N+2 briefing.	
	Coordinate with CI to initiate OPSEC plan; inspect battalion area for physical security violations.	
N+3:30	Present the intelligence estimate.	
N+3:45	Confirm transfer of company arms room keys to the rear detachment S2 before departing to staging area.	
	Verify security clearances of attachments through supporting S2.	
N+4:30	Sweep battalion area with CI personnel.	
N+5:00	Impose restrictions on incoming/ outgoing mail.	When given proper authority.
N+6:30	Issue OPORD at staging area.	
N+8:00	Ensure rear detachment is briefed on security.	
	Provide latest intelligence update to commander and staff. Have LNOs, DS MI CDR attend.	

NOTE: Modify times depending on mission requirements and SOPs.

Table 4-3. N-Hour critical actions list.

H-HOUR		ACTION	ACTION OFFICER(S)
	______	Mission Receipt	
1	______	WARNO--	CDR, S2/S3
-	______	Do IPB--	S2 (w/asst from ENGR, Terrain, Weather, etc)
2	______	Mission Analysis	
2	______	Brief Mission to CDR---------------------------	Coordinating Staff
-	______	Initial CDR's Guidance	
2	______	Do COAs---	CDR, S2/S3
3	______	Staff IPB & COAs-------------------------------	S2, S3
4	______	Do Staff Estimates-----------------------------	Unit Staff
-	______	COA Analysis (wargaming)--------------------	CDRs, Unit Staff
5	______	Select Best COA--------------------------------	S3, Selected Staff
6	______	Do DST---	S2/S3, FSO, ADA, ENGR, SIG, CHEM
5	______	Brief DST	
7	______	Brief Selected COA-----------------------------	Selected Staff
-	______	CDR's Decision	
7	______	WARNO Scouts/Patrols Recon---------------	S2/Asst S2/BICC officer/S3
8	______	Mission Brief	
8	______	OPORD Prep/Staff Coord---------------------	Unit Staff
8	______	Turn in OPORD Annex	
9	______	Proof OPORD	
10	______	Issue Targeting Overlay------------------------	FSO
11	______	Brief Scouts/Patrols Recon--------------------	S2/Asst S2/BICC officer/S3/ENGR/FSO/ADA
12	______	Execution Matrix---------------------------------	S3
13	______	BOS Synchronization Matrix-----------------	Unit Staff
14	______	Maps/Charts Prep-------------------------------	Unit Staff
14	______	CDR's Map Updated----------------------------	S2/S3
15	______	OPORD Rehearsal------------------------------	XO, S3
16	______	OPORD Briefing---------------------------------	XO
17	______	Initial Backbrief	
17	______	TF CDR Brief Higher HQ---------------------	CDR, S2, S3
17	______	Co/Bn CDR Backbrief	
18	______	Rehearsal---	Unit Staff, CDR
19	______	Final INTEL Update-----------------------------	S2
	______	LD/LC	

NOTE: These items are not listed in a definite order; some may not occur at all.

Table 4-4. Mission planning sequence.

S2 ACTIONS UPON RECEIPT OF A MISSION:

1. Receive and analyze mission:
   - Receive CDR's guidance and assist in development of CDR's PIR.
   - Understand CDR's intent.

2. Define the battlefield environment:
   - Identify AO and AI.
   - Identify battlefield characteristics that will influence friendly and threat operations.
   - Identify intelligence gaps.
   - Develop preliminary IR.
   - Identify specified and implied intelligence tasks:
     - **Specified:** From OPORD Annex B, R&S Appendix, tasks to subordinate units, MI task organization for combat, coordinating instructions, and collection plan.
     - **Implied:** Mission dependent METT-T.
       - **M—Mission** (CDR's intent, PIR, scheme of maneuver).
       - **E —Enemy** (OB and IPB to develop collection IR).
       - **T—Troops** (Organic and Higher Assets).
       - **T—Terrain and Weather** (R&S asset LOS, LOCs, obstacles, weather effects on system's collection ability).
       - **T—Time Available** (Plan, Move, Operate, Report).
   - Determine availability of organic/attached/supporting intelligence assets.

3. Request support/information from higher HQ if items/information are not available organically:
   - Maps/Imagery/sketches/blueprints of objectives, NAIs.
   - Collection plan, asset available times and asset tracks.
   - Weather - light data and climatic summary.
   - Enemy OB data.
   - Terrain products (LOC/vegetation/CCM/MCOO/elevation/LOS/hydrology).
   - Higher R&S plan.
   - Determine R&S assets you and higher HQ will control (intelligence architecture).
   - Determine availability, coordinate intelligence support from non-MI systems and organizations (AN/TPQ-36/37, OH-58D, FISTV, AN/TPS-25A, AN/TPS-58B, Engr, Co/teams, Cavalry, chemical units).
   - Obtain SOI; sign and countersign information through unit SIGO.
   - Advise higher of your recommended PIR and expected intelligence gaps.

4. Distribute maps, imagery, and sketches (maintain accountability).

5. Brief staff on abbreviated intelligence estimate to assist their mission planning.

6. Issue WARNO to R&S assets.

7. Describe the battlefield's effects:
   - Analyze terrain.
   - OCOKA factors.
   - Effect on friendly and enemy operations.
   - Analyze weather effects on terrain, troops, equipment, and friendly and enemy operations.

Table 4-4. Mission planning sequence (continued).

8. Assess other characteristics of the battlefield (mission dependent):
   - Politics.
   - Population.
   - Demographics.
   - Social ethnic/Religion.
   - Economic

9. Evaluate threat:
   - Evaluate enemy OB, current disposition and strength, committed and reinforcing units (maneuver, artillery, engineer, ADA, NBC, UW, air, intelligence assets, smoke, E-O sensors) and current significant activities.
   - Evaluate threat capabilities and tactics.
   - Develop/refine intelligence database and threat models.
   - Prepare doctrinal templates.
   - Prepare threat critical event lists.

10. Determine ECOAs:
    - Use doctrinal templates to develop event and situation templates (include all forces that can affect mission completion), AAs and MCs.
    - Identify enemy mission and intent.
    - Determine (in simple terms) how the enemy sees us acting in this operation.
    - Determine how the enemy is currently disposed and what must be accomplished to get from now to his desired end state, given that this ECOA must account for our actions determined in 3 above.
    - Determine enemy concept of the operation and subunit tasks. How will each of the following contribute to accomplishing the intent determined in 2 above:
      - Maneuver.
      - RISTA.
      - FS.
        - Artillery/mortars/rockets.
        - CAS.
        - Aviation.
        - NBC weapons.
        - EA.
      - Air defense - Recommend task organization for MI assets.
    - Identify most likely and most dangerous ECOAs.
    - Identify IR and NAIs for each friendly COA identified; TAIs with S3/FSO for each COA.
    - Develop HPTs for each friendly COA with S3/FSO.
    - Assist in developing HVTs for each ECOA.
    - Integrate EA into targeting.
    - Determine BDA criteria (damage required for each target, a means to determine damage).
    - Finalize and prioritize PIR (obtain CDR's approval).
    - Define indicators and SIR.
    - Send WARNO on threat to subordinate units to facilitate planning.

11. Develop ISM:
    - Tie to PIR/IR (include LTIOV).
    - Include timeline, required decisions, decision criteria, SIR, collection assets.
    - Plan for system cross-cue, back-up coverage.
    - Ensure it answers CDR's requirements for intelligence, targeting, and BDA in time to make decisions.

Table 4-4. Mission planning sequence (continued).

- Participate in staff wargame of DST and targeting plan (as opposing CDR).
- Never hold MI assets in reserve.

12. Develop R&S plan and overlay to support DST (with support from S3/FSO/ENGR/ ADO) (see **Reconnaissance Planning** below):
- Determine indicators, SIR, NAIs, DPs, TAIs.
- Ensure NAI/TAI support synchronization or R&S plan with fire plan and scheme of maneuver.
- Identify intelligence gaps and request collection support from higher (RII).
- Develop communications plan/reporting schedule.
- Wargame R&S and collection plan.
- Update ISM, R&S, and/or collection plan.
- Receive, consolidate, and deconflict subordinate R&S plans and overlays.
- Consider "Reconnaissance Fundamentals" [Maximum reconnaissance force forward, orient on the location or movement of the reconnaissance objective, report all information rapidly and accurately, retain freedom to maneuver, gain and maintain enemy contact, develop the situation rapidly (FM 17-95).]
- Consider "R&S Principles" [(1) Tell the commanders what they need to know in time for them to act and (2) Do as much as possible ahead of time (FM 34-2-1)].

RECONNAISSANCE PLANNING

While the DMP for reconnaissance operations is conducted in the same manner as for any other combat operation, the following steps will assist the reconnaissance planner in ensuring that the unique features of reconnaissance operations are addressed.

Reconnaissance Mission Analysis

■ Identify Reconnaissance Objectives. Reconnaissance mission analysis must identify the objectives of the unit's reconnaissance effort. These objectives are obtained from—
- PIR.
- DPs.
- HPTs/HVTs.
- Confirming Events. (These are enemy actions and counteractions that confirm an ECOA and are determined by event analysis.)

The reconnaissance planner may determine that some of the above requirements may be accomplished by RFIs to higher HQ, but all of those that are not must become reconnaissance objectives.

■ Identify Reconnaissance AO. Reconnaissance planners identify the area in which reconnaissance forces will be operating, either specified by control measures such as LOAs or CFLs or implied by the location of reconnaissance objectives.

■ Restate Reconnaissance Missions. Once reconnaissance objectives have been determined, the mission for each element of the unit reconnaissance force must be determined from task analysis and stated as a subunit task. The task analysis process—determining specified, implied, and essential reconnaissance tasks and identifying limitations and constraints—is described in FM 101-5. Reconnaissance missions will be determined to be reconnaissance-in-force or zone, area, route, or force-oriented reconnaissance missions.

■ Identify Available Reconnaissance Force. Reconnaissance planners must identify the subunits that are available to execute the unit's reconnaissance effort and their current and projected status.

Table 4-4. Mission planning sequence (continued).

■ Identify Available Equipment. Reconnaissance planners must identify the equipment available in an FMC status that will be used by elements of the reconnaissance force to accomplish reconnaissance tasks.

RCOA Development

■ Identify Required Reconnaissance Force. Reconnaissance planners must identify the force required to accomplish the unit's reconnaissance objectives. The number and type of elements involved will be based on the following:

- Number, type, location, and sequencing of reconnaissance objectives. Some objectives will require constant surveillance. Some will be time-phased and need only be observed for part of the battle. Some elements of the reconnaissance force will be able to accomplish several reconnaissance objectives.

- Type of reconnaissance mission. A route reconnaissance will require a different element than a reconnaissance-in-force.

- Strength, composition, and disposition of enemy security force. For each element of the reconnaissance force, the decision will have to be made whether it will have to infiltrate or penetrate the enemy security force. Infiltrations dictate stealthy elements and techniques while penetrations will require the correct reconnaissance force ratio.

■ Identify Required Equipment. Each reconnaissance objective will require specific reconnaissance equipment based on the following:

- Proximity of the reconnaissance element to the objective. What is the effective observation range?
- Environmental conditions. What will be the impact of severe weather? Night?
- Signature provided by the target. Does it emit a signal? Heat? Does it need to talk to anyone?

■ Determine the Reconnaissance Scheme. The RCOA must ensure that all reconnaissance objectives are addressed by elements of the reconnaissance force that can provide the read on the objective required by the commander. The reconnaissance force must therefore be in the right place, at the right time, and with the right observation, communications, and force protection equipment. The RCOA must deploy the reconnaissance force in depth to ensure that contact is not lost with moving reconnaissance targets. It must identify reconnaissance objectives that cannot be addressed so that RFIs may be prepared or risk accepted. In developing the RCOA, reconnaissance planners must answer the following questions:

- Are all PIR, HPT, DP, and confirming events addressed as reconnaissance objectives or RFI?
- What role will other BOS/combat operations play in the reconnaissance scheme?
    - FS (to include EA and SEAD for air insertions/extractions).
    - Air defense.
    - Mobility/counter-mobility.
    - CSS.
    - Deception.
    - OPSEC.
- How will this RCOA—
    - Maintain focus on the reconnaissance effort?
    - Gain and **maintain** contact with reconnaissance targets?
    - Provide early warning?
- How will we ensure that reconnaissance operations remain the bde/TF initial main effort?

Table 4-4. Mission planning sequence (continued).

- What is the reconnaissance force's relationship with security force?
- How far forward will we conduct reconnaissance? Is there a force in front of us? What are the control measures?
- How will fire support be integrated into this RCOA?

■ Determine Task Organization, C^2 Relationships. Once the reconnaissance force and its associated equipment are determined, it must be task organized to accomplish the RCOA in the most efficient manner. The C^2 structure must address unity of effort by identifying the COR. **The COR must have the control structure and authority to direct the reconnaissance force.** Reconnaissance operations often take place over large distances and long periods of time where communications systems are subject to attack. The communications plan will require careful attention to the use of retransmitted teams and redundant systems.

■ Identify Control Measures. While reconnaissance operations require the same positive control measures as any other, some require special attention.
- LOA. This measure identifies the forward edge of the reconnaissance battlespace and is usually associated with a CFL or FSCL.
- BHL. Reconnaissance forces operate throughout the unit's AO, but must be able to coordinate the acceptance of reconnaissance targets entering their battlespace with the unit to their front.
- TF Forward Boundary. In brigade operations there will often be a need to identify a brigade zone forward of the subordinate maneuver TFs.

■ Determine CSS Requirements:
- Resupply: How often, by whom, where, how?
- CasEvac: By whom, how?
- Reconstitution: From where will reconstituted reconnaissance forces come? What are the priorities?

RCOA Analysis and Wargaming

■ Key Events to Wargame During Analysis of an RCOA:
- Infiltration or penetration of the enemy security force.
- Insertion/extraction methods.
- Backup communications plan.
- CasEvac operations.
- Change in reconnaissance objectives based on new information.
- Handover of reconnaissance target leaving reconnaissance unit's LOS or AO.

R&S Ten Keys To Success

1. Commander's intent and **CCIR** drive R&S.
2. Support **R&S** early with integrated staff products.
3. Be adept at abbreviated **IPB** for **Quick** or **Combat Decision Making Process.**
4. S2 focus on the event template to capture the moving enemy.
5. S2 then provides **situation template,** depicting enemy in the engagement area.
6. Use automation tools like **ASAS** to enhance products.
7. Deploy organic R&S assets early and request support from higher. Issue S3 **FRAGO** with S2 graphic overlay.
8. Provide continuous coverage throughout the depth of the battlefield.
9. Use all possible assets from national level to scouts.
10. Evaluate reporting and provide **immediate and continuous feedback** to the commander.

APPENDIX A

UNIT INTELLIGENCE TRAINING CHECKLIST

This checklist will help a newly assigned brigade or battalion S2 maintain technical and tactical intelligence proficiency. Its building-block approach is systematic. The phases are supported by tasks to be completed for certification during the time specified. Some tasks may not apply to a particular echelon of staff.

PHASE I - ORIENTATION PHASE

This phase covers basic orientation to the division and the unit's mission. **This phase should be completed within the first 30 days**. The goal is to quickly acquaint the S2 with an understanding of the battalion, brigade, and division roles within corps, possible missions, and deployment areas.

Initial/Date	Task
__________	Coordinate with and receive briefing from outgoing S2.
__________	Review unit's intelligence library of publications listed in Appendix J.
__________	Read unit METL and latest training brief to determine overall unit missions and status.
__________	Receive unit CDR in-brief.
__________	Talk with unit CDR; ask what are his desires for intelligence and intelligence products. Tailor intelligence to the needs of the CDR.
__________	Visit and receive briefing from ACofS G2, Deputy G2, G2 Operations Chief, G2 Plans Chief, G2 Training Chief.
__________	Meet with subordinate S2's and gain an understanding of—
__________	• Their commander's intelligence priorities.

_______ • Personnel and equipment status.
_______ • Training proficiencies and deficiencies.
_______ • Areas where they need your echelon's help.

_______ Visit MI battalion or DS MI Co commander for unit and equipment briefings, display, and demonstrations.

_______ Study division, brigade, and battalion OPLANs.

_______ Review and evaluate intelligence annex to unit SOPs. (Use FM 34-1, FM 101-5, your SOP, and intelligence annex to the division SOP.) Provide CDR an updated evaluation of the unit intelligence annexes. Develop an SOP if one is missing or outdated.

_______ Receive an organizational and functional briefing from the next higher G2/S2.

_______ Study the CDR's current PIR and IR as developed.

_______ Meet and discuss intelligence missions, products, and training with the—

_______ • Scout platoon leader.
_______ • GSR platoon leader.
_______ • Local division CI team.
_______ • ACT.
_______ • ACE Chief, IPS Chief, and CM Chief.
_______ • EWO.
_______ • Terrain detachment team.

_______ Determine all organic or assigned division intelligence collection assets that may support collection efforts. (See Appendix G.)

_______ Learn to use section's automation equipment and programs.

________ Learn the connectivity, reporting, and requesting procedures.

________ Meet with next higher echelon G2/S2 to discuss the CM process.

________ Review unit's collection plan.

________ Become knowledgeable of the unit's and division's FSOP, RSOP, and deployment procedures.

________ Read unit and division history.

________ Obtain SI access through the SSO and review SCI billets for unit.

________ Review unit's arms room security SOP.

________ Inventory all sensitive items in one company.

________ Review key control program SOP.

________ Review physical security and crime prevention SOP.

________ Receive a crime prevention and physical security program briefing from the MP and Directorate of Security.

________ Become familiar with the S2 section vehicles and generators, as necessary.

________ Become familiar with the TOC SOP, setup, and multisystem, multiechelon connectivity.

________ Visit G2 training for intelligence training products.

________ Test an arms room's J-SIIDS.

________ Visit unit staff sections and attached sections from other units (FSO, ADA, NBC, SOCCE, and ENGR).

Initial/Date	Task
________	Develop and implement a detailed internal security inspection program that covers intelligence areas within the unit.
________	Develop working relationship with higher echelon G2/S2.
________	Review mission and section METL and battle tasks. Have section members give their assessment of current training level.

PHASE II - FUNDAMENTAL PHASE

This is the most critical phase for the newly assigned S2. The knowledge, insight, and assessments of I&S will establish the foundation for future development. This phase focuses on the IPB process, intelligence production and dissemination, collection operations, and security. **This phase should be completed within 90 days**.

Initial/Date	Task
________	Present to unit personnel an officer professional development class on the threat.
________	Update unit's collection plan as needed; submit revisions to the CDR and forward a copy to higher HQ G2/S2.
________	Brief the CDR on division and corps IEW assets.
________	Brief next lower units on division and corps IEW assets.
________	Inspect unit map program. Identify basic map load and training requirements.
________	Inspect and evaluate unit training for intelligence tasks, including awareness of AR 381-10.
________	Determine country and area study requirements from open and classified sources based on contingency and assigned OPLANs.

_______ Begin country studies.

_______ Understand additional duty responsibilities as required:

- _______ Security manager.
- _______ Physical security officer.
- _______ Personnel security officer.
- _______ Crime prevention officer.
- _______ Key custodian.
- _______ Top secret document custodian.
- _______ Information systems security officer.

_______ Receive briefing and training from unit SIGO on communications procedures and communications equipment PMCS.

_______ Become familiar with capabilities, limitations, and vulnerabilities of unit's equipment (tanks, Bradleys, other assigned, attached, DS systems, or OPCON BOS operating in the unit's battlespace; particularly the TA and information dissemination capabilities).

_______ Update, as needed, the unit's—

- _______ Arms room SOP.
- _______ Physical security SOP.
- _______ Key control SOP.
- _______ Crime prevention program.

_______ Conduct random unannounced security spot checks. Report the results and any recommendations to the CDR.

_______ Brief CDR on the unit's intelligence estimate.

_______ Brief CDR on the intelligence collection and reporting system from platoon through corps.

_______ Develop, update, and evaluate an S2 section METL and battle tasks.

______ Be knowledgeable on CTC threat vehicles, E-O devices, doctrine, and tactics. Teach a class on this subject at least once in this phase.

______ Visit nearest battle simulation center and become familiar with computer operations.

______ Visit and observe training of primary weapon systems in the division.

______ Understand current division policy on CONUS and OCONUS transportation of classified material.

______ Work with commanders. Ask what intelligence they want during FTXs and OPORD briefings.

______ Know how to read a TACFIRE direction system message. Let FSO know you want to receive these during FTXs and deployments.

______ Work with FSO to develop and review unit targeting and BDA procedures.

______ Know unit air LNO. Request BDA and in-flight reports that will bolster reporting.

______ Obtain recent CTC and applicable CALL. Develop plan to correct any deficiencies.

______ Obtain inspector general checklist and CIP checklists; conduct internal inspection on all S2 functional areas. Brief CDR on inspection results and solutions.

______ Review language proficiency requirements; assist S3 in coordinating training. Retain list of unit's capabilities; include unpaid abilities.

______ Conduct classified materials inventory; learn destruction and verification methods; review destruction and transfer SOPs.

__________ Constantly review clearance requirements, applications, and updates.

PHASE III - DEVELOPMENTAL PHASE

This phase will build and expand on the knowledge gained during Phase II. It will cover the same basic intelligence functions but in greater detail. **This phase should be completed within 180 days.**

Initial/Date	Task
__________	Fully understand the functions of the S2 at the unit's TOC and how S2 functions drive other TOC functions.
__________	Submit REDTRAIN requests through applicable higher S2 to G2 training, as required.
__________	Review threat OB holdings and files. Order update material as required.
__________	Determine available intelligence databases and actually "work the system" to obtain data in support of a CONPLAN. Brief CDR on the results.
__________	Identify gaps of information in battle books, OB holdings and files, and intelligence annexes. Submit RIIs to satisfy intelligence requirements.
__________	Identify S2 and intelligence weaknesses from the last EXEVAL and present a correction plan to the CDR.
__________	Evaluate intelligence training classes and provide written results to the CDR.
__________	Evaluate unit soldier's knowledge of intelligence training tasks and provide a written report to the CDR.
__________	Determine if all ADP systems are accredited. Brief CDR on steps needed to correct any problems.

Initial/Date	Task
__________	Demonstrate a thorough understanding of the intelligence reporting system during FTXs.
__________	Act as threat CDR during the wargaming process and TF rehearsals.
__________	Develop an internal S2 section intelligence cross-training program. Have section soldiers teach the classes.
__________	Determine unit's security posture during FTXs and recommend corrective actions to the CDR.
__________	Become familiar with and teach a class on doctrine, tactics, and current vehicles of a CONPLAN threat.

PHASE IV - ADVANCE PHASE

This phase will take the S2 who has mastered the intelligence fundamentals through **the first year,** during which the unit will have participated in the full annual training program.

Initial/Date	Task
__________	Successfully complete an EXEVAL.
__________	Successfully deploy section and perform S2 operations during a CTC rotation.
__________	Successfully deploy section and perform S2 operations during a BCTP Warfighter Exercise.
__________	Qualify or familiarize yourself on all unit weapon systems.
__________	Update unit's intelligence annex as required.
__________	Research a classified intelligence topic that is relevant to the unit's mission and brief the CDR on the conclusion.

_____ Accurately predict threat actions during FTXs using IPB and all intelligence assets.

_____ Successfully integrate with higher echelon G2/S2.

APPENDIX B

INTELLIGENCE PREPARATION OF THE BATTLEFIELD PRODUCTS

This following examples of various overlays, templates, and matrices are constructed and used during the IPB process to enhance the battlespace visualization of the commander and staff. These samplings are not all-inclusive. Your products will be dictated by your mission statement, resources available to you, and the time allowed.

Figures B-1 through B-6 are terrain overlays. Figures B-7 through B-13 are templates and associated matrices. Figure B-14 is a sample of an ADA-tailored IPB.

The MCOO includes—

- **MCs** and **AAs** stated in terms of concealment and cover, size and deployment type of formation that can move along them (armored battalion in company wedge), speed the formation can reasonably achieve; dismounted, mounted, and aerial routes; trails and passes that will allow column movement if unopposed.
- **Obstacles** in terms of what they are an obstacle to.
- **Key terrain**.
- **Restricted terrain** in terms of what size force can deploy there and what mobility work could be done to"unrestrict" it.
- **Severely restricted** terrain in terms of what will allow a dismounted force to traverse it and at what speed.
- **IVLs**. Without a 1- or 5-meter resolution terrain product, IVLs will have to be determined by historical knowledge or ground reconnaissance.
- **Soils** and **weather analysis** that indicate trafficability and"dig-ability."

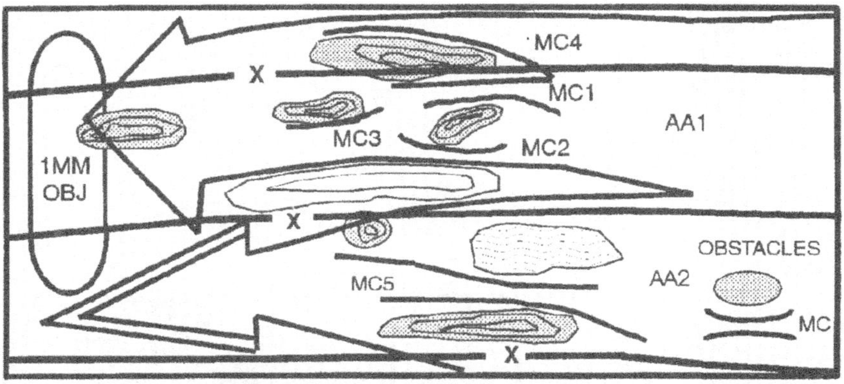

Figure B-1. Modified combined obstacles overlay.

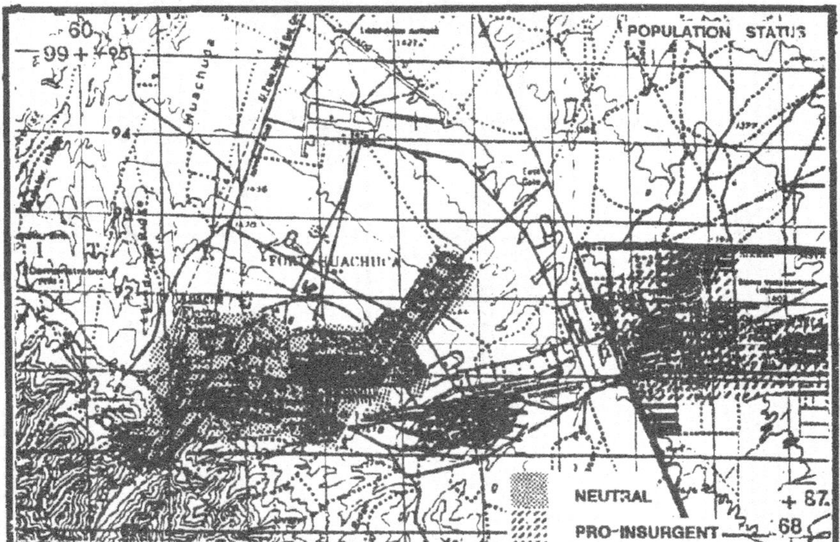

Figure B-2. Population status overlay.

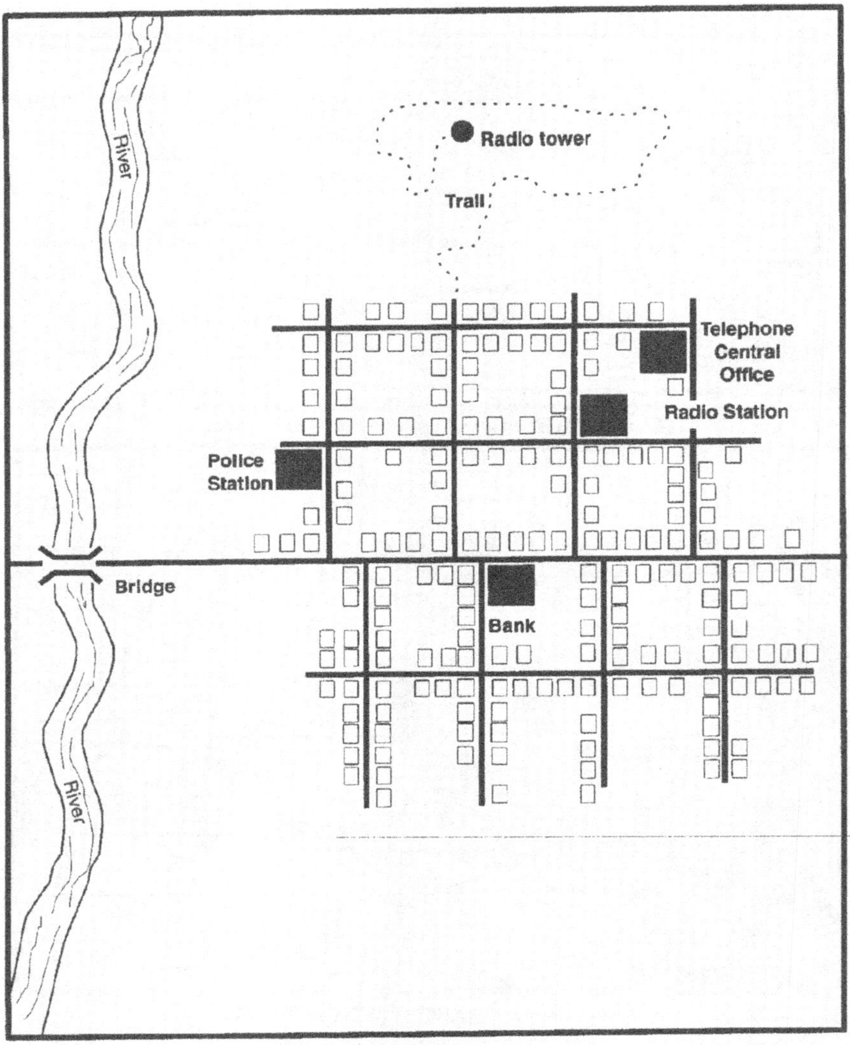

Figure B-3. Key facilities and target overlay.

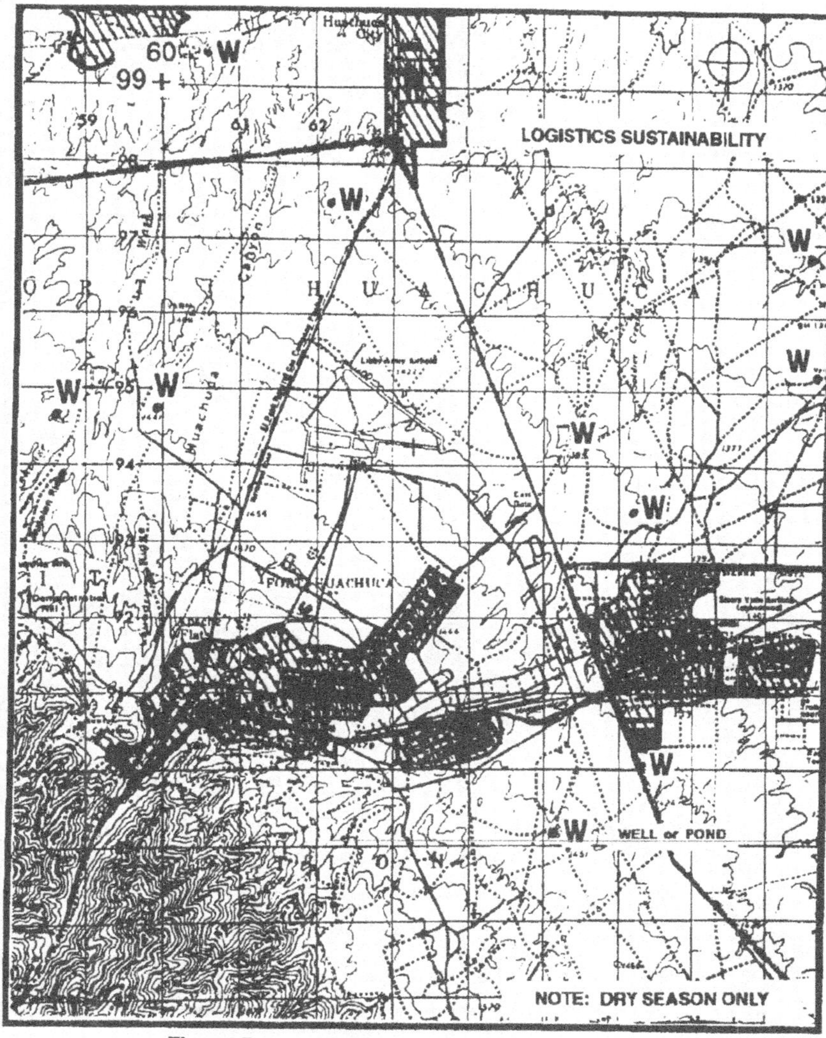

Figure B-4. Logistics sustainability overlay.

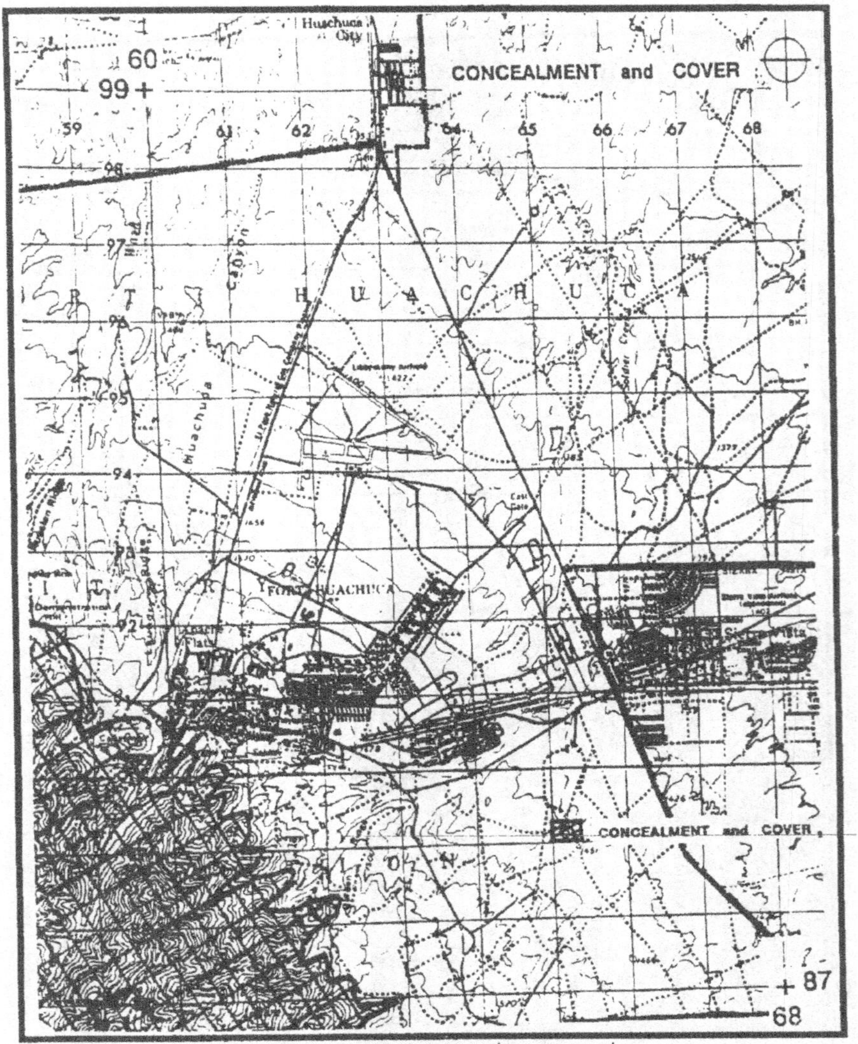

Figure B-5. Concealment and cover overlay.

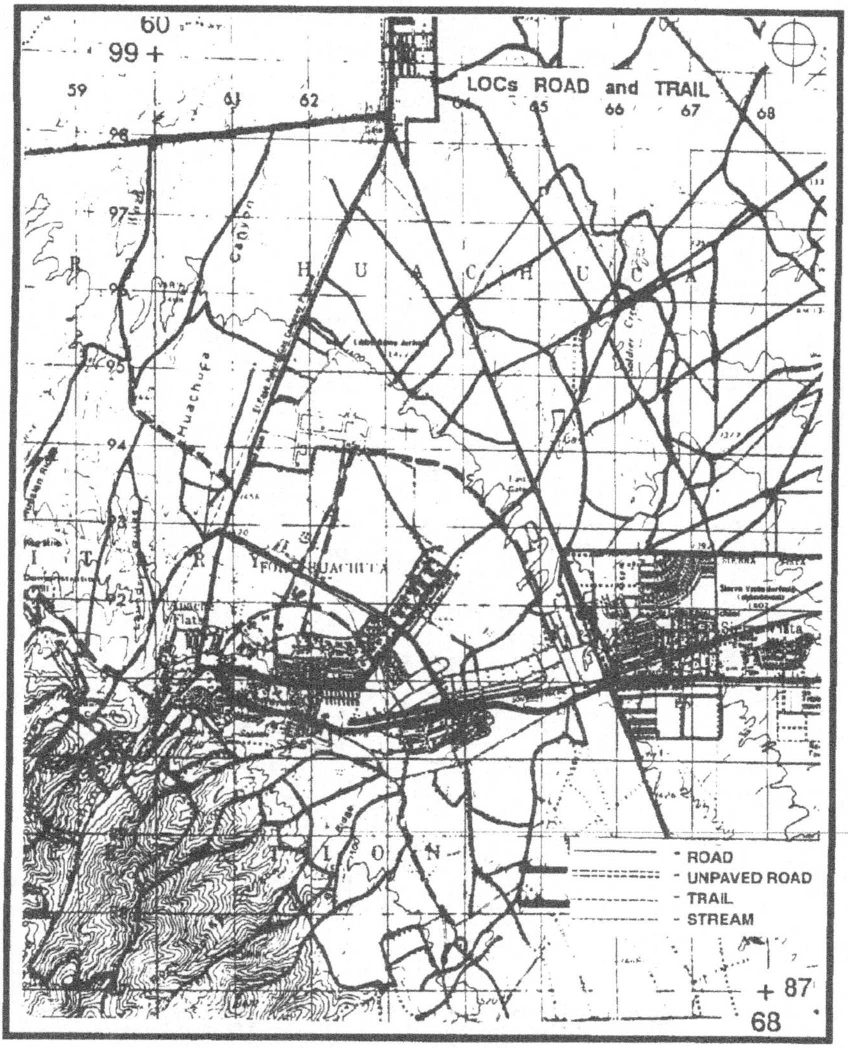

Figure B-6. Lines of communication overlay.

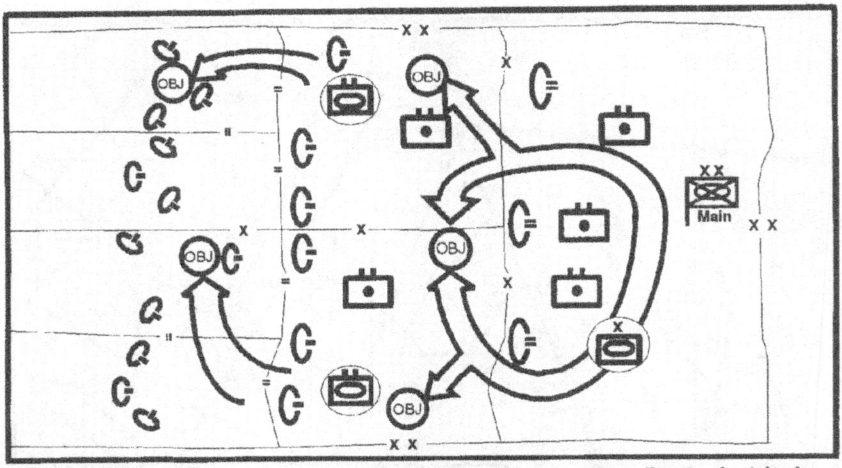

Figure B-7. Doctrinal template - depicts enemy forces according to doctrinal deployment, unconstrained by terrain.

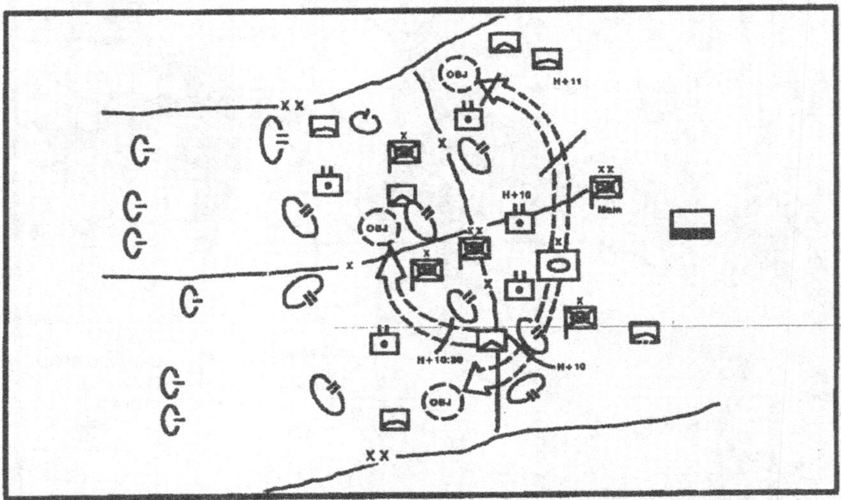

Figure B-8. Situation template - depicts deployed enemy forces adjusted for obstacles and terrain.

The situation template should include the following for mechanized armor-based offense and defense.

In the offense:	In the defense:
• Objectives & LOA. • AAs & MCs. • Recon routes, OP & IEW sites. • Firing lines & direct fire range fans. • Formations & deployment lines. • Artillery targets & range fans. • Artillery & ADA position areas. • ADA coverage. • Attack helicopter routes, BPs & range fans. • LZs. • CAS routes. • Situational obstacles. • Chemical agent targets. • Smoke targets	• BPs & direct fire range fans. • Reserve & hide positions. • CATK routes & firing lines. • Counter-SOPs & counterrecon forces. • Recon routes, OP & IEW sites. • Ambush sites. • Artillery targets & range fans. • Artillery & ADA position areas. • ADA coverage. • Attack helicopter routes, BPs & range fans. • Tactical & protective obstacles. • CAS routes. • Situational obstacles. • Chemical agent targets. • Smoke targets.

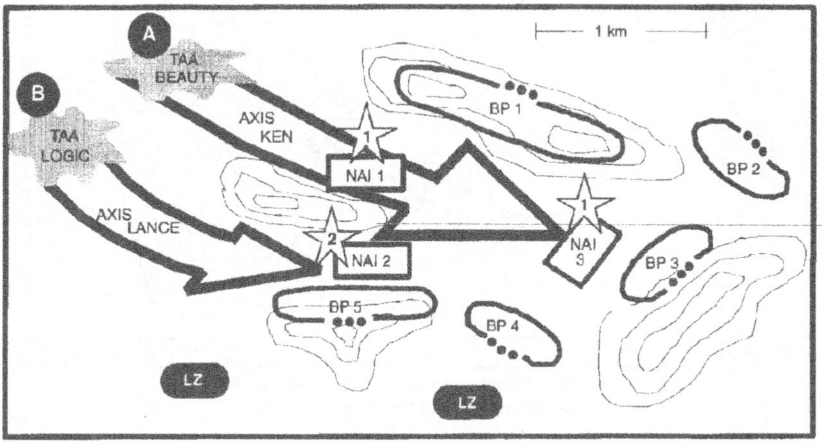

Figure B-9. Decision support template.

DP No.	1	2
Decision Criteria	Insurgent Camp is in NAI 1 or 3.	Insurgent Camp is in NAI 2.
Maneuver	A Co receives 3/B, occupy TAA BEAUTY, O/O movement to CATK along AXIS KEN.	A Co occupies TAA BEAUTY, O/O occupy BPs 1, 2, and 3.
FS	Priority: A, B, C.	Priority: C, B, A.
M-CM-S	1/A/13th Engr to A.	1/A/13th Engr to C.

Figure B-10. Partial BOS synchronization matrix.

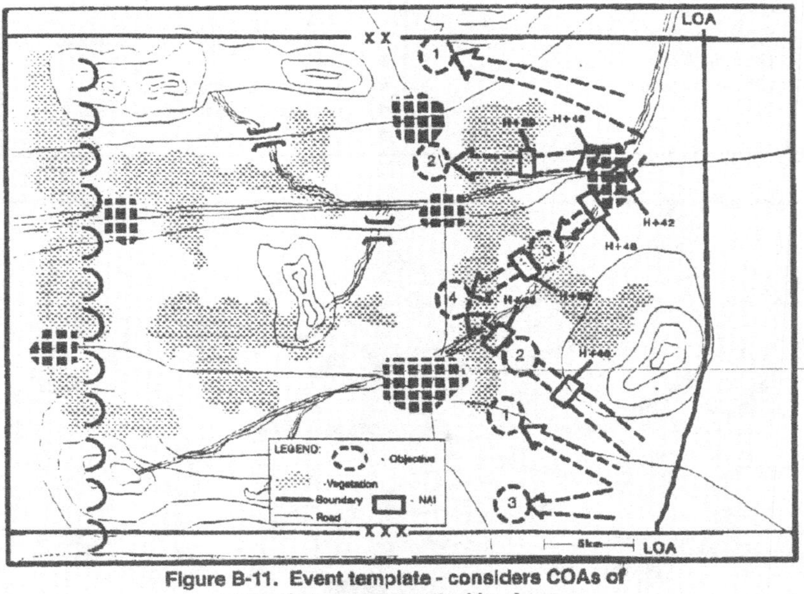

Figure B-11. Event template - considers COAs of reinforcing or counterattacking forces.

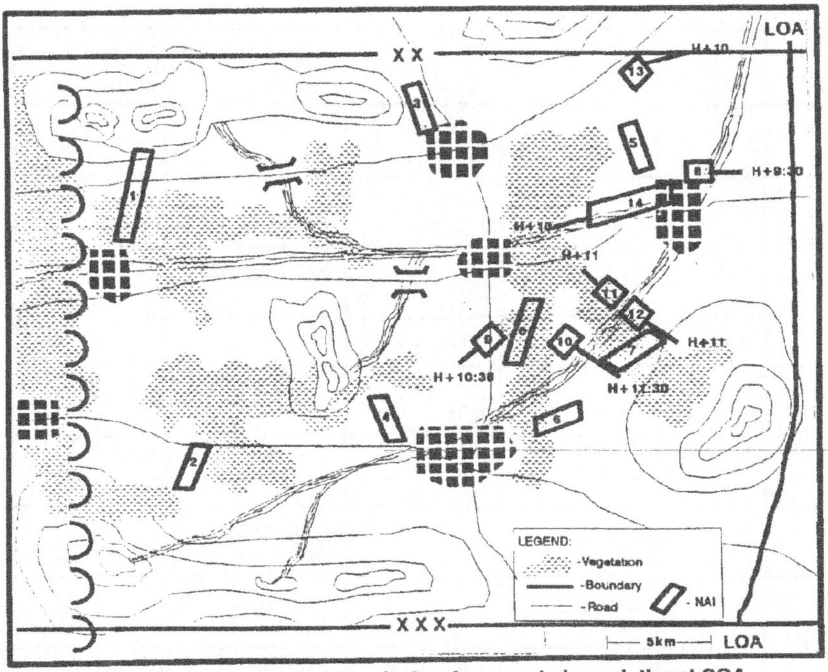

Figure B-12. Event template - depicts key events in each threat COA.

NAI	EST TIME	INDICATORS THAT WOULD HELP CONFIRM COA 1	COA 2	COA 3	OTHER
1	H-15		AASLT Forces		
2	H-15		AASLT Forces		
3	H-15	AASLT Forces		ASSLT Forces	
4	H-15		Infil of LT Inf		
5	H-4		LT Inf ATK NAI 5		
6	H-4	LT Inf ATK NAI 7	LT Inf ATK NAI 6		
7	H-4	LT Inf ATK NAI 8		LT Inf ATK NAI 7	
8	H-4	■Poised to ATK		LT Inf ATK NAI 8	
9	H-4	■Poised to ATK	■Poised to ATK		TD ATKs
10	H-4		■Poised to ATK	■Poised to ATK Abreast	TD ATKs
11	H-6			■Shifts N	
12	H-18				1 or 2 Bdes ATK N
13	H-18				1 or 2 Bdes ATK S
■ Indicates ECOAs.					

Figure B-13. Event matrix.

CONSIDERATIONS FOR AIR IPB

1. DEFINE THE BATTLEFIELD .

- AI for air IPB is significantly larger than ground IPB.
  - Includes airfields which can range supported unit AO to include those within aerial refueling radius.
  - AAAs which lead into supported unit's AO.
  - Three types of air avenues:
    * F/W Attack = Normally follows ground forces.
    * F/W Transport = Bombers and R/W lift uses terrain to mask ingress and egress routes.
    * R/W Attack = Normally follows ground forces.

2. DESCRIBE THE BATTLEFIELD'S EFFECTS . See the following example of weather effects on aviation operations.

TYPE AIRCRAFT	MINIMUM VIS	WIND-SPEED & DIRECTION[1]	PRECIPITATION	TEMPERATURE & HUMIDITY[2]	ILLUMINATION
Fixed Wing	2.5 nmi vis 2.5 nmi AGL ceiling in hilly terrain 1.0 nmi AGL in flat terrain	For airborne operations speed must be <13 knots	Severe weather within 3 miles of target will hinder acquisition. Freezing rain greatly limits lift capability	Temp >100 and humidity <80% will degrade payload capability	Little air-to-air or point CAS capability at night except newest F/W
Rotary Wing	1.0 nmi vis 300 ft AGL	For airmobile operations speed must be <13 knots	Same	Same	No CAS without illumination except newest R/W

[1] Strong winds perpendicular to AAA increase difficulty in hitting target.
[2] Amount of degradation differs with type of aircraft.

Figure B-14. Sample of an ADA-tailored IPB.

- Terrain Analysis.
  - Determine impact of geographic factors on the ability of the aircraft to approach, acquire, and engage a target.
  - Predict how an aircraft would most likely approach target or area to deliver ordnance.
  - Locate positions that rotary wing could use for stand-off and/or pop-up attacks.
  - Determine locations of possible LZs, DZs, and the AAAs to these areas.
  - AAAs do not stop at FLOT or FEBA and they do not enter your AO solely from the FLOT or FEBA; watch your flanks and rear.
  - The closer an aircraft is to a target the greater an influence terrain will have on the aircraft and how it is employed.
  - The best method for formulating AAAs is based on fixed factors:
    * Locations of enemy airbase or staging area (known or suspected).
    * Position of friendly targets.
    * Most aircraft will fly a straight-in approach from the airbase to the vicinity of the target.

    Reason: Shorter overall mission time, less fuel needed, more payload or ordnance.
  - Exceptions which cause deviation from this rule:
    * Heavy concentrations of air defense.
    * Major terrain features that can mask the approach of aircraft at low altitudes.
    * Terrain that may force unnecessary exposure.
    * Terrain exceeding the operating ceiling of rotary-wing aircraft.
  - What happens in the vicinity of the target?
    * Most attacking aircraft will drop lower to use terrain masking and minimize early warning and exposure times to associated air defenses.
    * How will the enemy aircraft approach the target (Hi-Hi-Lo, Lo-Hi-Lo, or Lo-Lo-Hi, etc)?
    * To engage the target the pilot must be able to see it (visual, radar, sensor).
    * The pilot must approach his target so he can see it at far enough distance to allow lock-on and engagement by his weapon system.
    * The pilot will maximize his standoff range to increase survivability.
  - How to determine the direction of attack:
    * Know the enemy HPT; what friendly asset he must destroy.
    * Know the location of these assets.
    * Know the location of known or suspected enemy airbases and FARP.

Figure B-14. Sample of an ADA-tailored IPB (continued).

* Determine AAA which could be used by fixed- or rotary-wing aircraft making the attack.
* Determine locations from which stand-off and pop-up rotary-wing attacks are most likely.
* These locations will provide terrain masking for the aircraft to hide behind.
* The location must also provide observation and fields of fire into the target area when the aircraft pops-up.
* This type of location + a masked ingress and egress route = a stand-off or pop-up position.
* Within the vicinity of the target, consider the following:
  _ LOS acquisition of the target.
  – Weapon stand-off capability.
  – Pilot reaction time.
  – Aircraft survivability.
  – Standard tactics.

3. EVALUATE THE THREAT.

● Should address the following:
– AOB—How many and what type aircraft. (Include different models such as Mi-24 E or F because they all have varying payloads, optics, and weapon systems.)
– Capabilities of the aircraft in the enemy inventory.
– Types of ordnance available to the enemy.
  * Guns.
  * Rockets.
  * Tactical ASMs, CBUs, PGM, AAM, NBC, and mines.
– Capabilities:
  * Range.
  * Guidance (aerial, ground).
  * Release altitude.
– Doctrine:
  * Raid sizes and composition.
  * Ingress and egress altitude and speed.
  * Delivery profiles.
  * Use of EP and SEAD.
– C^2:
  * Most non-Western aircraft operate under strict ground control. They are vectored toward a target rather than being sent to an area to seek

Figure B-14. Sample of an ADA-tailored IPB (continued).

targets on their own. The vector is from ground emplaced beacons which normally operate in HF range.

- Work to template high value C^2 targets:
  * GCI nodes.
  * FAC.
  * RNP.
  * Beacons.
- Maintenance and Sortie Generation Capability:
  * OR rates determine the percentage of aircraft able to fly. OR rate x number of aircraft = the maximum aircraft that can fly at any one time.
  * The average sortie generation rate for most modern air forces is 2.5 to 3 for first 24-hour period. Second and succeeding day's planning factor is 5% of available number of aircraft.
  * To get sortie generation rate: number of aircraft x sortie generation rate = number of sorties in a 24-hour period.
  * To estimate number of sorties in your AO, consider ground situation, enemy main effort, enemy doctrine.

4. DETERMINE THREAT COURSES OF ACTION.

- Enemy air operations are conducted in support of their ground operations. The most important part of determining enemy air COAs is to understand the ground situation and enemy ground COAs.

- With the information and analysis you've done in the first three steps, determine how the enemy will employ his air assets.

- Consider enemy doctrine, AAAs, terrain in the target area, and current weather conditions.

- You must answer the following questions:
  - When will the enemy commit his air assets?
  - Where will the enemy commit his air assets (target area)?
  - What air assets will the enemy commit (number of aircraft)?
  - How will the aircraft attack in the target area (AAA or delivery profile)?

Figure B-14. Sample of an ADA-tailored IPB (continued).

APPENDIX C

RECONNAISSANCE AND SURVEILLANCE PLANNING

R&S planning is continuous. You must understand the commander's intent for each mission. After mission analysis, the commander should tell you the key information needed before and during the mission. This key information is the commander's PIR and IR. The PIR and IR provide the initial focus to R&S planning. Plan to conduct R&S throughout the depth of the battle in space and time. (See FM 34-2-1.)

G2/S2 R&S RESPONSIBILITIES (Table C-1)

Table C-1. G2/S2 R&S responsibilities.

REQUIREMENT	EFFECTS
Coordinate with higher, flank, forward units.	Better guidance from higher & integration with adjacent units; less redundancy & gaps in coverage.
Identify R&S requirements; rehearse R&S plan.	Success of unit mission.
Supervise & coordinate the command's aerial & ground reconnaissance (include LRRPs).	Better use of survl systems (GSR/REMBASS employment), LRRPs, and tasking UAVs or other aerial assets to answer CDR's PIR.
Brief GSR, REMBASS, and patrol team leaders (plt leaders of every BOS in AO to collect) (e.g., LLVI, LRSD, scouts).	Effectiveness of R&S assets which, in turn, affects success of R&S effort.
Prepare R&S WARNO and OPORD early-on.	Start the reconnaissance effort ASAP.
Update R&S OPORD as info is received.	Effectiveness of R&S effort and unit mission.
Provide R&S plan to higher, lower, and adjacent units.	R&S effectiveness and other unit actions; prevention of negative effects.

EXAMPLE: The primary R&S planning objective is to task subordinate units (all possible assets; e.g., MI, MP, FA, Signal, Engr, FSB) to cover NAIs important to brigade mission or directed by higher HQ. Subordinate units plan the use of available R&S assets to satisfy taskings from the brigade and their own requirements. They strive to complete their plans quickly (using the 1/3, 2/3 rule) so R&S assets have time to prepare and execute the plan. As soon as they are complete, the subordinate unit plans are consolidated at brigade to form the brigade R&S plan. You now have a draft plan that shows when and in what areas to begin R&S planning.

R&S PLANNING PROCESS (Figure C-1)

I. PLANNING PROCESS.

A. Initial Requirements:

1. Did higher HQ provide tasking requirements?
2. Was the commander's PIR and IR stated and included?
3. Did commander provide R&S intent?
4. Did S2 brief the staff on enemy collection capabilities?
5. Were other staff tasks performed?

B. SIR Developed (IPB Driven):

1. Did S2 identify air or ground AA?
2. Did situation and event templates reflect probable or prioritized ECOAs?
3. Was NAI developed in detail? (What is expected? When? Where?)
4. Were collectible indicators at NAIs developed?
5. Were SIR developed from NAIs and indicators?
6. Were reporting requirements developed for priority collection missions to allow the commander time to change plans?

C. Possible Collectors Analyzed:

1. Did S2 coordinate with staff, S2, and G2 to identify all available collection assets?
2. Did S2 analyze asset capabilities to develop collection requirements based on range to target, time available, target characteristics, terrain, weather, enemy (obscurant use), and communication?
3. Did S2 analyze collection redundancy (is it necessary)?

Figure C-1. R&S planning process.

4. Did staff identify support requirements (communication nets, retransmission, FS, NBC reconnaissance support, logistic support, special equipment support)?
5. Did S2 identify gaps in collection?
6. Did S2 back brief S3 or the commander on R&S concept?
7. Were WARNOs sent to appropriate assets?
8. What were timelines?
   a. When was mission received?
   b. What is NLT time for execution?
   c. When was templating done?
   d. When was tentative plan made? Back briefed?
   f. When was initial reporting needed?
   g. Who was in charge of R&D planning?
   h. Who was in charge of counterreconnaissance planning?
   i. What is LTIOV.

II. PREPARATION FOR R&S OPERATIONS.

A. Specific Collection Instructions:

1. What assets could be available? Used? **(See Figure C-4.)**
   a. Scouts.
   b. Patrols.
   c. Forward Observers.
   d. Armor.
   e. Aviation.
   f. Infantry.
   g. EW.
   h. REMBASS.
   i. Chemical.
   j. UAV.
   k. Medical.
   l. CI teams.
   m. GSR.
   n. OPs and LPs.
   o. Signal.
   p. Antitank.
   q. Engineer.
   r. Cavalry.
   s. MP.
   t. ADA.
   u. SOF.
   v. Support units.
   w. GRCS.
   x. Interrogation teams.

2. Did the S2 provide detailed instructions to tasked assets? Did the instructions include—
   a. Who is tasked?
   b. What to look for?
   c. Why to look?
   d. When to look?
   e. Where to look?
   f. What you could expect to see?
   g. How to get there?
   h. Who to coordinate with?

Figure C-1. R&S planning process (continued).

i. Passage of lines and recognition signals?
j. Reporting requirements?
k. Friendly assets in AO?
l. Resupply?
m. Obstacles?
n. MEDEVAC?
o. NBC hazards?

3. Was the collection location appropriate (concealment, collectable)?
4. Were there sufficient control measures included to control assets during mission?
5. Did the S2 request assistance from higher HQ for identified collection gaps?
6. Did the R&S plan cover all collection requirements?
7. Were assets overtasked?
8. Was redundancy appropriate for this mission?

B. Coordination:

1. What is format of plan (collection plan, overlay, matrix)?
2. Were direct or indirect fires or jamming coordinated between staff and S2?
3. Was a CI vulnerability assessment made?
4. Was additional equipment (special) planned for?
5. Were communication nets established to meet reporting needs?
6. Were commanders and staff briefed on plan before execution?
7. Did commander or S3 approve final plan?
8. Did assets know specific requirements (PIR and IR)?
9. Was plan disseminated to all involved or with a need to know (FRAGO, overlay)?
10. Was plan sent to higher HQ?

C. Asset Internal Coordination:
1. Was equipment checked?
2. Were internal procedures clarified?
3. Did coordination between assets occur?
4. Was mission rehearsed?
5. Was plan developed far enough in advance for assets to prepare or rehearse?
6. Was plan developed in time for higher HQ to review?

III. EXECUTION.

A. Continuity of R&S and counterreconnaissance operations:

Figure C-1. R&S planning process (continued).

1. Did unit plan provide for operations when scout or other R&S assets are inoperable?
2. Did unit SOP provide for operations during briefings, debriefings, or rehearsals?
3. Are unit and leaders cross-trained to facilitate substitutions or replacement of scouts?

B. Asset and Unit Response:

1. Did assets depart and set up on time?
2. Did assets use concealment, cover, and camouflage?
3. Were assets able to observe enemy undetected?
4. Was low-level deception used?
5. What were meteorological report requirements?
6. Were enemy locations pinpointed?
7. Was objective reconnoitered?
8. Were obstacles identified and marked?
9. Were routes marked?
10. Did assets submit route/obstacle overlay?
11. Was enemy reconnaissance located?
12. Were counterreconniassance missions performed?
13. Did assets help with C^2 during attack?
14. Did assets help direct or control fires?
15. Was terrain reconnoitered (trafficability reported)?

C. Reporting:

1. Were reports timely, accurate, and concise?
2. Were assets debriefed?

D. Results:

1. Did S2 plot asset reports (track results of plan)?
2. Did S2 identify inadequately tasked or unproductive assets and change tasking (with approval of commander and S3)?
3. Did reports or analysis answer PIR and IR?
4. Was R&S plan updated and recoordinated?
5. Were templates updated?

E. Dissemination:

1. Was commander briefed on answer to PIR?
2. Has commander updated PIR?
3. Did units get intelligence based on priority?
4. Did higher HQ get answers to taskings?
5. Did assets receive feedback on level of success?

Figure C-1. R&S planning process (continued).

OTHER R&S PLANNING CONSIDERATIONS (Figure C-2)

1. General.
   a. Combat information is knowledge of the enemy, weather, and terrain.
   b. R&S collection should answer the commander's PIR and IR.
   c. Every soldier is an important HUMINT collection asset.

2. Analysis of information collected on PIR and IR is an objective process.

3. Determine requirements:
   a. Set priorities. Develop SIR and SOR. Do not let the R&S elements decide what answers the commander's PIR.
   b. Terrain and weather.
   c. Tactical situation (deep, close, rear).
   d. Focus the R&S mission.

4. Develop R&S plan based on the principles of reconnaissance:
   a. Timely information.
   b. Aggressive reconnaissance.
   c. Continuous reconnaissance.
   d. Focus combat power.
   e. Relevant information (focused on PIR and CCIR).
   f. Reconnaissance OPSEC. Coordinate with G3/S3 to establish control measures which deconflict and help protect R&S assets on the battlefield.
   g. Accurate information.
   h. Complementary with operations.
   i. Consideration of threat counterreconnaissance threat.
   j. Destruction of threat reconnaissance elements when detected.
   k. Plan for CasEvac, resupply, and reconstitution for teams in the field.

5. Identify METT-T.

6. Develop communications plan and reporting schedule.

7. Issue WARNO to R&S asset leaders.

8. Coordinate R&S plan with—
   a. Commander and commanders of areas in which teams will operate.
   b. Adjacent and higher G2/S2.

Figure C-2. R&S planning considerations.

c. Unit S1/S3/S4.
d. FSO.
e. SIGO.
f. Engineer LNO.
g. ADA LNO.
h. Chemical Officer.
i. G3/S3 to tie R&S plan into the maneuver plan.

9. Brief R&S team leaders and all other R&S assets.

10. Advise higher and lower S2 of R&S plan.

11. Request support from higher headquarters.

12. Write R&S tasks into paragraph 3 of the OPORD.

13. Evaluate reports, review requirements, and identify intelligence gaps.

14. Analyze reports to deny or confirm your templates; then update templates.

15. Ensure combat information is passed immediately to those elements that can react to it.

16. Update the R&S plan based on known intelligence.

17. Plan for future operations.

18. Maximize use of all available assets.

Figure C-2. R&S planning considerations (continued).

You can now begin adding some detail to the R&S plan. Integrate any requirements from higher HQ. Translate the initial PIR and IR into indicators on which R&S assets can collect. Use the indicators to develop the SIR and SOR. The SIR and SOR ensure assets are collecting specific information that answers the PIR and IR.

Once you have a picture of the coverage required for the R&S effort, prioritize the SIR. Assign a high priority to those SIR that, when answered, will provide the greatest amount of intelligence in the shortest amount of time. After you prioritize the SIR, draft the R&S plan and distribute to the appropriate R&S assets. It is critical that you deploy R&S assets early-on and update them as additional information is received.

TOOLS TO TRACK DEPLOYED R&S ASSETS

Three methods are used to track deployed R&S assets:

- Reconnaissance base OPORD (Figure C-3).
- R&S overlay (Figure C-4).
- R&S tasking matrix (Figure C-5).

Select any method as long as the assets receive clear, specific instructions. Some S2s use all three methods by attaching the R&S overlay and R&S tasking matrix to the reconnaissance order prior to dissemination. When possible, using all three is best because each asset then has a graphic, a matrix, and a written set of instructions.

RECONNAISSANCE ORDER:

The reconnaissance order is the narrative instructions for deployed assets. Figure C-3 is an example of a brigade reconnaissance order.

RECONNAISSANCE ORDER

1. SITUATION.
   a. Enemy forces.
   b. Friendly forces.
   c. Attachments and detachments.
   d. Assumptions (OPLAN only).

2. MISSION. Expressed as task and purpose derived from planning process. Clearly stated reconnaissance objectives.

3. EXECUTION. Commander's intent, derived during planning process:

a. Concept of Operation. State how reconnaissance assets are operating in relation to the parent organization. This scheme must identify if the reconnaissance force is infiltrating or penetrating the enemy's security forces (or both) and how this will occur.

Figure C-3. Example of a brigade reconnaissance OPORD.

(1) Manuever.
(2) FS. (Refer to FS annex if required.)
(a) Scheme of fires that specifies what fires are available and how to get them.
(b) Engagement criteria.
(c) Scheduled fires in support of insertions and extractions.
(3) Mobility and survivability.
(4) Air defense.
(5) C^2W.

b. Tasks to maneuver units. [List all units from task organization into subparagraphs, and only unique items for those units; all other tasks go into d below.] For each of the elements in the reconnaissance force, state the following:
(1) How to get there.
(2) Where to look (NAI for static targets).
(3) What to look for.
(4) When to look for it.
(5) Plan to maintain contact with moving target.
(6) What equipment to bring.
(7) How (what communications systems, how often) to report.
(8) To whom to report:
(a) Brigade reconnaissance teams.
(b) Task forces.
(c) MI company (EW, UAV, other).
(d) ADA battery.
(e) Engineer battalion.
(f) FA battalion.
(g) Chemical company.
(h) MP platoon.
(i) Signal company.
(j) FSB.

c. Tasks to combat support units (if applicable).
(1) Fires (or refer to FS annex).
(a) Air support.
(b) Chemical support.
(c) FA support.
(2) Mobility and survivability. (Refer to mobility and survivability annex if required.)

Figure C-3. Example of a brigade reconnaissance OPORD (continued).

(3) Air defense. (Refer to air defense annex if required.)
(4) C^2W. (Refer to C^2W annex if required; address applicable subcomponents of C^2W: deception, EW, psychological warfare, UAVs.)

d. Coordinating instructions.
(1) Time or condition when OPORD or OPLAN becomes effective.
(2) EEFI.
(a) PIR.
(b) EEFI.
(c) FFIR.
(3) Risk reduction control measures.
(4) Environmental considerations.
(5) Any additional coordinating instructions.
(a) LD or defend NLT ____ .
(b) MOPP level.
(c) Rally points, alternate positions.
(d) Rehearsal time, type, location.
(e) Extraction plan.
(f) PZ and LZ.
(g) Fratricide avoidance measures.
(h) ROE and actions on contact.
(i) IHL.

4. SERVICE SUPPORT. State the concept of logistic support.
a. Support concept.
b. Material and services (resupply location or time and days of supply, emergency resupply, vehicle evacuation).
c. Medical evacuation (CasEvac plan).
d. Personnel (reconstitution plan).
e. Civil-military cooperation. NA.
f. Miscellaneous.

5. COMMAND AND SIGNAL.
a. Command (C^2 structure, report plan).
b. Signal (nets used, retransmission or relay plan).

Figure C-3. Example of a brigade reconnaissance OPORD (continued).

R&S OVERLAY:

The R&S overlay is the R&S plan in graphic format. Its purpose is to show exactly where and when R&S assets are operating. Use FM 101-5-1 for overlay graphics and symbols. Due to various R&S operational techniques and the unique nature of some SASO missions, you will need to construct some new graphics and explain them in your legend (Figure C-4).

The R&S overlay has two parts: (1) The graphic display of deployed or planned deployment of R&S assets, and (2) The marginal data consisting of the legend, administrative data, specific instructions to each asset, and the distribution list. The latter part consists of standard wording found on all overlays. (See top of Figure C-4.)

With digitized systems coming online, the location, heading, range to target, system health, and logistical data will be available NRT in digital, preformatted report format available to the S2/S3/S4 in the TOC, TAC, and ALOC.

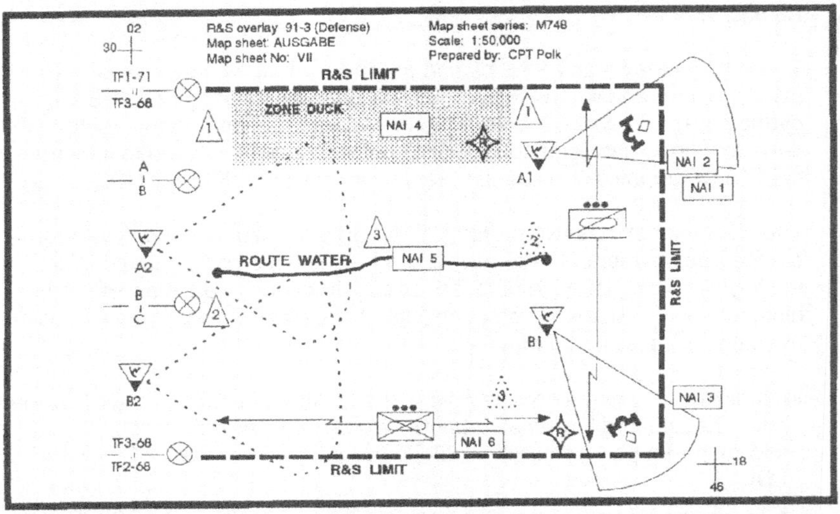

INSTRUCTIONS:

SCOUTS: Establish forward screen NLT 1600 at coordinate_________. Target enemy recon (BRDM, BMP, tanks). Coordinate with B Co for route to screen position and GSR team.

REMBASS: Establish two strings vicinity____and____ NLT 1800. Target enemy dismount recon and MR troops infiltrating along forest trails. Coordinate passage with B Co, Scout platoon, and GSR team.

GSR: Establish position A1 vicinity____ NLT 1700. Target enemy recon (BDRM, BMP, tanks, dismounted troops). Coordinate with A Co for passage, Scout platoon, and REMBASS team. Establish position A2 on order. Coordinate with B Co for position. Establish position B1 vicinity____ NLT 1700. Target enemy recon. Coordinate with B Co for passage, Scout platoon, and REMBASS team. Establish position B2 on order. Coordinate with C Co for position.

PATROLS: A Co conduct recon and security patrols in Zone Duck. Begin patrols NLT 1830 to 0600. Target enemy dismounted troops, BMP, and BRDMs. Coordinate with GSR team and Scout platoon. C Co establish screen along southern flank NLT 1800. Target enemy recon, BMP, BRDM, tanks. Coordinate with Scout platoon and GSR team.

OP/LP: Position as per battalion R&S SOP.

REPORTS: Report on battalion intelligence net using SALUTE format. Patrol leaders report on intelligence net after each patrol. Report combat information on intelligence or operations net. Assets report using schedule in R&S SOP. React to jamming as per battalion TAC SOP.

DISTRIBUTION: A, B, and C Companies, Scout platoon, GSR team, REMBASS team, S3, FSO, and brigade S2.

Figure C-4. R&S overlay with instructions on acetate.

The key portion of the administrative data contains the written instructions to each R&S asset. The instructions should focus on—

- The required operational timeframes. Give each asset a start and a finish time for each observation or action.
- The target. To answer the PIR, look for specific indicators. Each asset should be told exactly what to look for. Never give broad-based generic mission statements. Specific guidance will promote specific answers.
- Coordinating instructions. All assets at one time or another will move through or near another unit's AO. To keep units from shooting (directly or indirectly) friendly R&S assets, assets and units must coordinate with each other. FSO should restrict fires around R&S assets.
- Reporting requirements. All assets should know when, how often, and what format to use when reporting. Provide frequencies, alternate frequencies, and reactions during jamming, as well as the LTIOV for specific information to be reported.

Initially, the locations for assets are areas in which you recommend that they deploy. After the assets have deployed to the NAI, they report to the S2 or the S3 the actual coordinates from which they can best accomplish their mission. The S2 or the S3 then updates the R&S overlay graphics to show actual locations. Include the following control measures on your R&S overlay:

- Friendly boundaries, R&S limit of responsibility, NAIs, SPs, RPs, and checkpoints.
- Graphics depicting route, area, and zone reconnaissance.
- Primary, alternate, and supplementary positions.
- Sectors of scan for sensors.

R&S TASKING MATRIX:

Another less preferred method of disseminating R&S guidance is the matrix method (Figure C-5).

- **Priority.** The priority number of each mission, which should correspond with the PIR number.
- **NAI.** The NAI number and grid coordinate.
- **START/STOP.** The times for this mission.
- **SIR INSTRUCTIONS.** Explains to the assets exactly what they are looking for (target).
- **TASKINGS.** Lists the actual assets tasked to conduct each mission. An "X" placed under each asset identifies the tasking.
- **COORDINATION.** Tells the assets which units to coordinate with for this mission.
- **REPORTS.** Provides the assets with reporting requirements.

(Classification)

DTG: ______________

MISSION: ______________

A	B	C	D	E	F	G	H	I	J	K	L	M	N	O	P	Q
PRIORITY	NAI	START / STOP	SIR/INSTRUCTIONS	A	B	C	D			SCOUTS	MORTARS	GSR	REMBASS	OTHER	COORDINATION	REPORTS
1	1 2 3	1800 / 2400	When? Where? What type equipment does the recon element have? Do not Engage							X					With B Co for route to screen position. With GSR TM and REMBASS TM.	Per BN R&S SOP.

(Classification)

Figure C-5. R&S tasking matrix.

BRIEFING OPORDs

The following OPORDs are additions to the base reconnaissance OPORD (see Figure C-3). GSR, REMBASS, and patrol employment OPORDs would be issued to the teams by the respective platoon leaders.

GSR EMPLOYMENT:

- Give GSR teams adequate time to prepare and deploy.
- Deploy GSRs for best area coverage.
- Deploy GSRs, with a security force, so they scan an area and provide early warning.
- Plan logistical support.
- Limitations:
  - — Limited to LOS acquisition and gives off detectable signal.
  - — Visibility hindered by heavy precipitation, thick vegetation, and broken terrain.
  - — Limited range and ability to identify except by personnel and wheeled or tracked vehicles.

- Brief team leader using the OPORD format at Figure C-6.

1. SITUATION.

   a. Enemy. Intelligence estimate format.

   b. Friendly.
      (1) Current unit locations.
      (2) Scheme of maneuver.
      (3) Boundaries.
      (4) Objectives.

2. MISSION. Who, what, where, when, and why for this R&S mission. Surveillance of threat antiaircraft from location and time.

3. EXECUTION.

   a. Concept of operation.
      (1) Primary and alternate routes to sites (SPs, PPs, checkpoints, RPs).
      (2) Primary, alternate, and supplementary sites.
      (3) Sectors of scan.
      (4) Operating times.

   b. Coordinating instructions.
      (1) Battalion and brigade R&S plan: patrols, scout routes, times, sizes, LP and OP locations, ground and air cavalry screens, REMBASS, asset replacement plan, NAI redundancy coverage backup plan, etc.
      (2) Link-up plan with security forces.
      (3) Obtain target list from FSO (recommend additional targets).
      (4) Coordinate fire control measures with FSO and plan NFA for team locations.
      (5) Obtain obstacle plan from TF engineer and check routes for obstacles.
      (6) Coordinate with supported commander to tie the R&S site into unit defensive plan or scheme of maneuver.
      (7) Review engagement and disengagement criteria.
      (8) Review actions upon mission completion.

Figure C-6. Example of a GSR OPORD.

4. SERVICE SUPPORT. Coordinate with S4, battalion maintenance officer, SIGO for—

a. Class I (Food).
b. Class III (Fuel).
c. Class IV (Barrier materials) (during prepared defense).
d. Class V (Ammunition).
e. Class IX (Repair parts).
f. Organizational maintenance.

5. COMMAND AND SIGNAL.

a. Signal.
(1) Issue SOI extract.
(2) Review reporting requirements and communications connectivity.
(3) Identify multiple reporting paths and methods.
(4) Review communications check requirements.
(5) Execute COMMEX prior to employment.
(6) Review actions upon EA.

b. Command.
(1) Review chain of command.
(2) Review leaders' locations.

6. SAFETY.

Figure C-6. Example of a GSR OPORD (continued).

IREMBASS EMPLOYMENT:

Brief team leader using the reconnaissance OPORD format (see Figure C-3) with the changes shown in Figure C-7.

1. SITUATION.

2. MISSION. Who, what, where, when, and why for this R&S mission. Monitor NAIs, DPs, routes.

3. EXECUTION.

a. Concept of operation.
(1) SIR required.
(2) Team leader recommends sensor mix, emplacement sites, and relay sites.
(3) Primary and alternate routes and transportation to emplacement sites, PPs, and checkpoints.
(4) Emplacement method.

b. Coordinating instructions.
(1) Battalion and brigade R&S plan: patrols, scout routes, times, sizes, LP and OP locations, ground and air cavalry screens, IREMBASS.
(2) Link-up plan with security forces.
(3) Obtain target list from FSO (recommend additional targets).
(4) Coordinate fire control measures with FSO.
(5) Obtain obstacle plan from TF engineer and check routes for obstacles.
(6) Review engagement and disengagement criteria.
(7) Actions upon mission completion.

Figure C-7. Sample IREMBASS employment OPORD.

PATROL EMPLOYMENT:

Brief patrol leader using reconnaissance OPORD format (see Figure C-3) with the changes shown in Figure C-8.

1. SITUATION.

2. MISSION.

3. EXECUTION.

    a. Concept of operation.
        (1) Maneuver.
        (2) Fire support.

    b. Subunit tasks.

    c. Coordinating instructions.
        (1) Actions at the objective.
        (2) Time of departure and return.
        (3) Movement techniques and order of movement.
        (4) Routes.
        (5) PPs.
        (6) RPs and actions at RPs.
        (7) Actions upon enemy contact.
        (8) Actions at danger areas.
        (9) Actions at halts.
        (10) Locations of friendly obstacles, mine fields.
        (11) Rehearsals and inspections.
        (12) Debriefings.
        (13) SIR.

Figure C-8. Sample patrol employment OPORD.

DEBRIEFING Debrief patrols using the sample debriefing format below.

1. Size and composition of team. Team leader, assistant team leader, radio operator, and observers.

2. Mission.

3. SIR.

4. Reporting requirements.

5. Time of departure, method of infiltration, and point of departure.

6. Enemy observed enroute.
   a. Ground activity.
   b. Air activity.
   c. Miscellaneous activity.

7. Routes.
   a. Dismounted.
   b. Vehicle.
   c. Air.

8. Terrain.
   a. Key terrain.
   b. Significant terrain.
   c. Decisive terrain.
   d. Terrain compartment.
   e. Terrain corridor.
   f. Map corrections.

9. Enemy forces and installations.

10. Miscellaneous information.
    a. Lack of animals or strange animal behavior. Did animal reaction to patrol's presence generate alarm? If so, when, where, type of animal, activity that caused the alarm, and nature of alarm.

Figure C-9. Sample debrief format.

b. Mutilated plants.
c. Strange, uncommon insects.
d. Abandoned military equipment, number and type.
   (1) Out of fuel.
   (2) Unserviceable; estimate reason.
   (3) Destroyed or damaged; how.
   (4) Operational equipment left intact.

e. Abandoned towns or villages.

11. Results of encounters with enemy forces and local populace.

12. Condition of team including disposition of dead and wounded.

13. All of team's maps or any other identifiable material returned with team. For missing items, describe item and approximate location lost.

14. Conclusions and recommendations.

15. Process CEE, CEM, photographs, and videos taken by teams.

16. Time of exfiltration, method of exfiltration, exfiltration point.

17. Routes of return.
   a. Dismounted.
   b. Flight.

18. Time of return.

19. Point of return.

20. Enemy observed on return route.

21. What pre-mission intelligence support was provided and not needed? What intelligence support was not provided but needed? What intelligence was provided but turned out to be misleading or erroneous?

Figure C-9. Sample debrief format (continued).

R&S TEN KEYS TO SUCCESS:

1. Know that the commander's intent or CCIR drives R&S.

2. Support R&S early with integrated staff products.

3. Be adept at abbreviated IPB forquick or combat decisionmaking process.

4. Focus on the event template to capture the moving enemy.

5. Provide situation template depicting enemy in the engagement area.

6. Use automation tools like ASAS to enhance products.

7. Deploy organic R&S assets early and request support from higher HQ. Issue S3 FRAGO with S2 graphic overlay.

8. Provide continuous coverage throughout the depth of the battlefield.

9. Use all possible assets from national levels to scouts.

10. Evaluate reporting and provide immediate and continuous feedback to the commander.

APPENDIX D

PRIORITY INTELLIGENCE REQUIREMENTS, INFORMATION REQUIREMENTS, AND INDICATORS

PIR are intelligence requirements associated with a decision that will affect the overall success of the command's mission. PIR may change in order or be replaced entirely.

IR are requirements for intelligence to fill a gap in the command's knowledge and understanding of the battlefield or threat forces. They target specific COAs.

Indicators are positive or negative evidence of threat activity or any AO characteristic which points toward threat vulnerabilities or adoption or rejection by the threat of a particular capability, or which may influence the commander's selection of a COA.

During wargaming, the S2 develops a set of PIR for each friendly COA. Each is linked to a specific enemy action that requires a friendly response (DPs).

PIR are those IR critical to accomplishing the mission. Wargaming will dictate which IR will become PIR as the mission runs its course.

The S2 should nominate PIR for approval ONLY FROM THE LIST OF ALREADY PLANNED AND COORDINATED IR.

EXAMPLE OF A POOR PIR:

"Will the enemy attack? If so, how, when, where, and in what strength?"

- The PIR contains five significantly different questions. Which one is the priority? Unless the S2 gives more guidance, the individual collector must determine which part of this PIR to work.

- The S2 probably knows more about the situation than "the enemy mightattack somehow, sometime, somewhere, and in some strength." The PIR as stated might prompt some collection assets to collect known information.

- Even for issues that the S2 does not know, the enemy can only select from a limited range of COAs due, for example, to terrain, weather, policies. If the PIR consider IPB, the S2, through more specific tasking, will minimize the chance that collection assets will look for the enemy where they are not likely to be.

- When the staff wargames, it may find some aspects of this question to be irrelevant to the present situation. For example, the defense may be fully capable of defeating an enemy attack regardless of when they actually attack. Why waste collection assets on a question that does not need to be answered?

EXAMPLE OF A GOOD PIR:

There is no "set" of PIR we can present that will be useful for all tactical situations any more than there is a set of maneuver paragraphs that the S3 can plug into any OPORD. Below are examples of *types* of PIR the S2 presents to the commander for approval. Because intelligence needs change over time, most PIR will be important only during certain times.

Referring to PIR as "timephased" is redundant. They are dynamic just as the battle will be dynamic. Just as there are no standard situation templates or friendly COAs that will serve in all situations, there is no standard set of PIR. Good PIR, however, have some things in common:

- They ask only one question.
- They focus on a specific fact, event, or activity.
- They provide intelligence required to support a single decision.
- They are tied to key decisions that the commander has to make.
- They give an LTIOV.

Examples:

- "Will the enemy defend OBJ JAMES using a forward slope defense?"
- What size force is defending OBJ NACO?"
- Which bridges over the San Pedro River are intact?"

Not all PIR are appropriate as questions, particularly those associated with moving targets or formations. PIR are *requirements* and should be stated as such when appropriate. In addition, PIR for moving targets must include the requirement to continue to maintain contact with the target. Simply locating the target once will not prevent it from surprising the commander later.

Example: "Locate and track the enemy's battalion reserve."

APPENDIX E

GRAPHIC REPORT FORMATS

Intelligence personnel use a variety of formats to request, report, and disseminate information (see FM 34-3). Personnel working in a joint environment should use JCS Pub 6-04.11. The formats assimilate information into automated intelligence support systems, such as the ASAS.

This appendix concentrates on graphic report formats, specifically the graphic INTSUM and the graphic intelligence estimate. The INTSUM is the main focus, but a graphic intelligence estimate is also provided.

GRAPHIC INTELLIGENCE SUMMARY

The graphic INTSUM is used by units at all echelons to rapidly convey threat situation dynamics. The graphic INTSUM is a current depiction of significant threat dispositions, activities, strengths, and weaknesses and an assessment of the most probable ECOA. A current, complete graphic INTSUM will provide for efficient collection and enable the commander to exploit fleeting threat vulnerabilities.

The graphic INTSUM can be maintained on conventional maps of any scale as an overlay or created electronically and disseminated to computer and television screens. For example, army-level organizations may utilize a 1:250,000 map scale and focus on an army or front-level threat, whereas in SASO activities the deployed unit may utilize a map scale of 1:25,000 or smaller and focus on a platoon or company-size threat.

The graphic INTSUM (Figure E-1) may contain 12 components:

- Legend and margin information.
- Area orientation.
- Threat unit locations and mission activities.
- Threat mission capabilities assessment.
- Threat boundaries and front-line trace.
- Threat objectives.
- Threat air activity.
- Threat weaknesses and vulnerabilities.
- Threat strengths and capabilities.
- Threat intentions assessment.

- Friendly commander's intent and PIR.
- Predicted future threat activity.

LEGEND AND MARGIN INFORMATION:

Components include—

- Overall classification of the INTSUM, posted top and bottom, to include a "Classified By" line and declassification instructions.

- Title.

- Issuing headquarters.

- "As Of" DTG. Use either local or Zulu time; this is critical since predicted threat activity will be depicted based upon the "As Of" DTG plus a given number of hours.

- Map reference, to include scale, and at least two coordinate registration marks.

- A symbology key can be added for increased clarity and**must** be included if symbols used are not in FM 101-5-1.

- SASO symbology is not standardized, but there are several symbol methodologies that have been developed. (See MIPB Jan-Mar 94, p 35.)

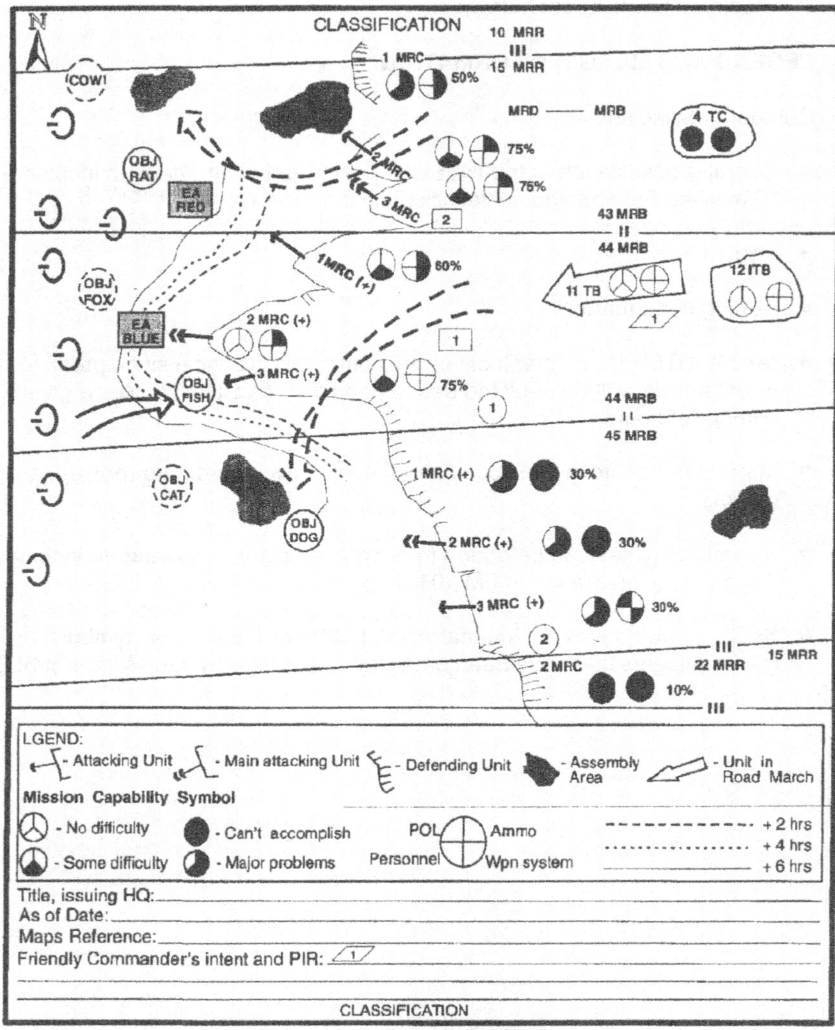

Figure E-1. Graphic INTSUM.

AREA ORIENTATION:

Selected geographic features will be highlighted on the overlay in order to orient the user to the area being displayed. The features selected will depend on the specific AO and type mission in which the friendly unit is involved.

Common features could include cities, rivers, and major relief features. Selected features for some SASO activities may be highlighted on a smaller scale. In an NEO highlighted features could include airports, LOCs, embassies, consulates, and American housing areas, and other features that will play prominently in the operation. Provide only the features necessary for orientation; do not clutter the INTSUM.

THREAT UNIT LOCATIONS AND MISSION ACTIVITIES:

Depict **committed** threat unit locations with a bracket followed by the unit identification. The size of the bracket will indicate the frontage occupied by the unit. The mission of the unit will be shown graphically (Figure E-2).

Depict **uncommitted** threat units with the symbology shown in Figure E-3.

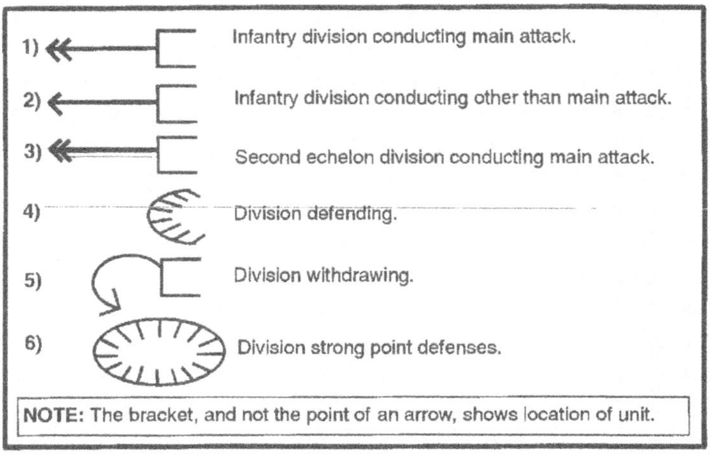

Figure E-2. Threat unit locations and mission activities (committed).

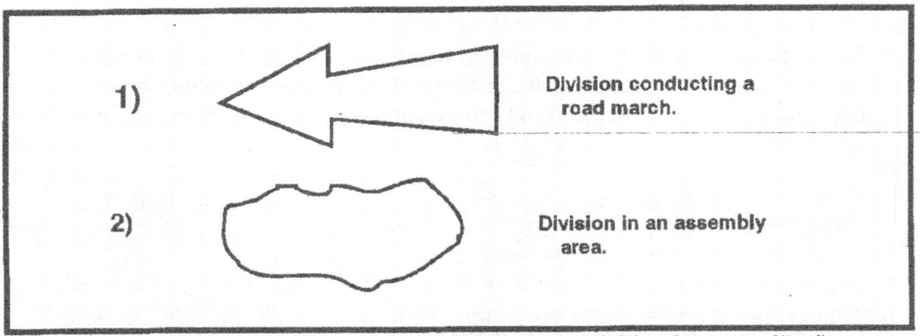

Figure E-3. Threat unit locations and mission activities (uncommitted).

THREAT MISSION CAPABILITIES ASSESSMENT:

Use combat effectiveness graphics to assess and display the capability of a threat unit to perform its mission. Consider all factors, tangible and intangible, in making the assessment; include unit strength, logistics status, maintenance and readiness status, morale, mission, weather, and terrain. Place selected symbols in the vicinity of the unit identification (Figure E-4).

1)  NO PROBLEM accomplishing mission.
80%>

2)  SOME DIFFICULTY accomplishing mission.
50% - 80%

3)  MAJOR PROBLEMS accomplishing mission.
41% - 50%

4)  CANNOT accomplish mission.
40%<

Figure E-4. Threat mission capabilities assessment.

A diminished threat mission capability is due usually to one or more specific problem areas such as damaged or destroyed weapons systems or a lack of ammunition, POL, or replacement personnel. If known, graphically depict these problem areas for each unit immediately next to the threat mission capabilities assessment symbol so the friendly commander can exploit what may be a very crucial, but fleeting, threat vulnerability.

Specific problem areas are indicated by blackening in the appropriate quadrant of the circle. The problem area categories are not static; other problem areas can be selected for display on this symbol as long as the selection is clearly explained in the key. Figure E-5 shows a sample problem area symbol and an example of the problem area symbol used in conjunction with the threat mission capabilities assessment symbol.

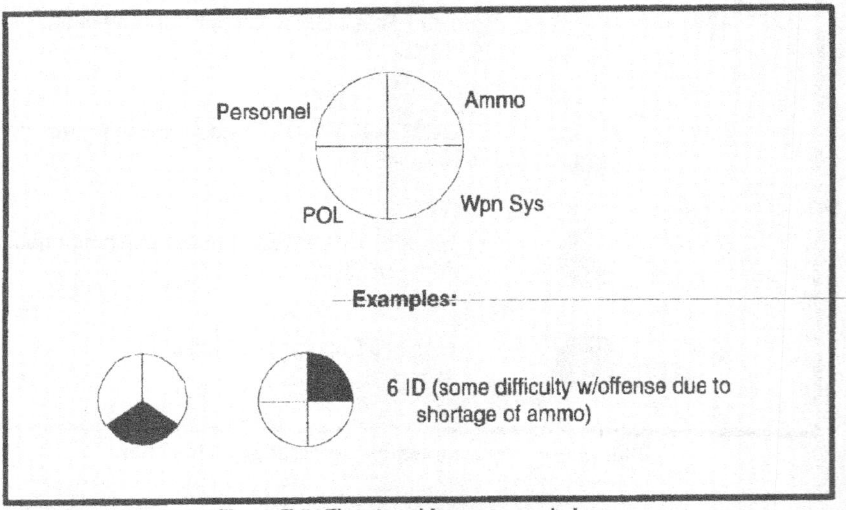

Figure E-5. Threat problem area symbology.

THREAT BOUNDARIES AND FRONT-LINE TRACE:

In conventional military operations, threat boundaries below the division level normally will not be included in the INTSUM. These lower level boundaries are often difficult to delineate and clutter the INTSUM. In some SASO activities, such as support to insurgency and COIN, smaller units will be the rule and must be constantly accounted for on the INTSUM.

The front-line trace will not be shown on the graphic INTSUM. The brackets depicting the frontages of the threat units in contact will suffice for an effective front-line trace. Addition of a front-line trace can be easily confused with the depictions of predicted threat activity.

THREAT OBJECTIVES:

If threat objectives can be discerned, they will be placed on the INTSUM.

THREAT AIR ACTIVITY:

Depict threat air activity on the INTSUM. Use an aircraft symbol with direction of travel and total number of sorties. Some units may opt to break down the sortie count by category, such as fighter (F), bomber (B), and reconnaissance (R). Example: A total sortie count of 12 could be depicted as (12). The same sortie count could also be depicted as (4F/6B/2R). The sortie breakdown should be used only if it will not clutter the INTSUM unnecessarily. Sortie breakdown symbols, such as F, B, and R must also be included (Figure E-6).

T

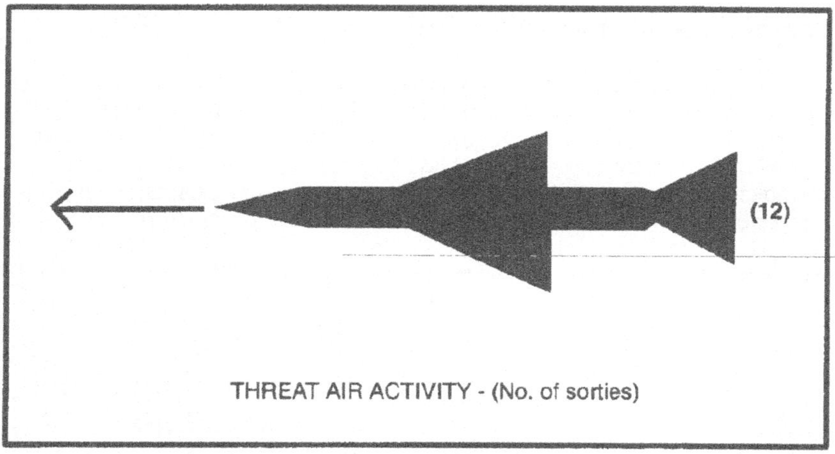

Figure E-6. Threat air activity symbology.

HREAT WEAKNESSES AND VULNERABILITIES:

Show specific weaknesses or vulnerabilities by using a numbered circle on the INTSUM (see Figure E-1). A supporting message could list threat weaknesses and vulnerabilities in narrative form.

THREAT STRENGTHS AND CAPABILITIES:

Show threat strengths and capabilities by using numbered squares (see Figure E-1).

THREAT INTENTIONS ASSESSMENT:

Write an assessment of threat intentions and most probable COA in a concise narrative format; graphically key it to the INTSUM when possible.

FRIENDLY COMMANDER'S INTENT AND PIR:

List the friendly commander's intent and PIR in narrative format on the INTSUM. Graphically key these statements to the overlay.

PREDICTED FUTURE THREAT ACTIVITY:

Draw predicted lines of threat advance or withdrawal and tie them to a timeline to convey as much information about future threat activity as possible (Figure E-7). Predicted lines of advance or withdrawal will be based on the "As Of" time found in the legend and margin information. Remember: Predicted lines are only an aid to planners; current threat situation must be continuously monitored to update the predictions.

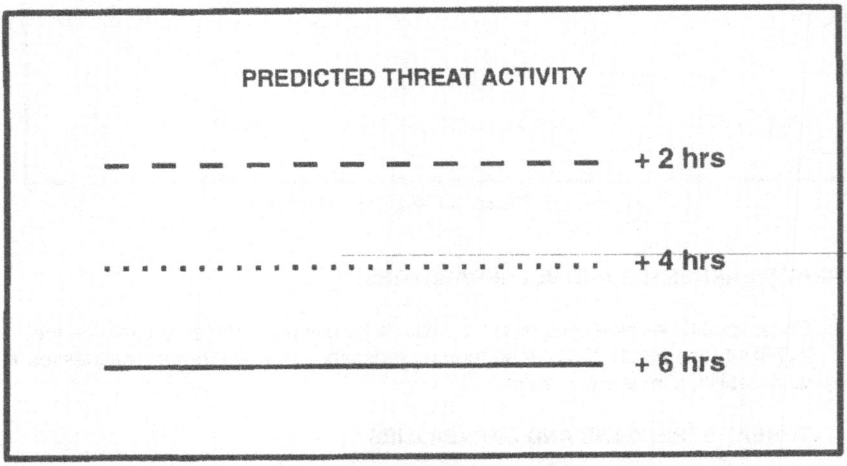

Figure E-7. Predicted threat activity timelines.

APPENDIX F

TARGETING

TARGETING PROCESS

Targeting is the process of selecting targets and matching the appropriate response to them, including operational requirements and capabilities. The purpose of targeting is to disrupt, delay, or limit threat interference with friendly COAs; it requires coordinated interaction between FS operations, intelligence, EW, and plans cells. Targeting is based on the enemy's assets that provide him an advantage (HPTs), friendly scheme of maneuver, and tactical plans. (See FM 6-20-10.)

Target development is one of the six IEW tasks. Mission analysis must precede generating any ECOAs. These ECOAs are the starting point for all other staff actions to wargame against, and to develop follow-on products. The products of the decide function are briefed to the commander. Upon his approval, his decisions are translated into the OPORD with annexes. (See FM 34-130.)

METHODOLOGY

Decisions create the guidelines for the acquisition and engagement of targets. TA and attack are made through a decision cycle. The methodology used to translate the commander's intent into a plan is**DECIDE, DETECT, DELIVER,** and **ASSESS.** Figure F-1 shows the functions of intelligence support to targeting, Figure F-2 shows the targeting methodology, and Figure F-3 is a targeting checklist.

RECEIVE GUIDANCE ON:	• Commander's intent. • HPTs. • Attack criteria. • Lead time between DPs and TAIs. • ROE. • Conditions to establish BDA requirements.
DEVELOP:	• MCOO. • Situation and event templates. • HVTs.
EXPLAIN:	• ECOA as part of wargaming. Based on friendly COA, refine event template; assist in developing HPT list, TSS matrix, and sensor or attack systems matrix.
PRODUCE:	• Collection plan, ISM, and AGM. • SOR.
BRIEF: (Ensure all analysts & collectors understand the CDR's intent.)	• ACE sections (ECOA, HPTL, TSS, and AGM).
COLLECT:	• Information for nomination, validation, and PSA.
DISSEMINATE:	• HPT related information and intelligence to FS immediately. • Pertinent information and BDA per SOP and TTP.
ENSURE:	• Information collection and intelligence production support all FRAGOs.

Figure F-1. Functions of intelligence support to targeting.

DECIDE	DETECT	DELIVER	ASSESS
• Target Development • TVA • HPT and HVT • TSS • Attack Options • Attack Guidance	• Target Detection Means • Detection Procedures • Target Tracking	• Attack • Planned Targets • Targets of Opportunity • Desired Effects • Attack Systems	• Tactical Level • Operational Level • Restrike • Feedback
Tracking your target is applicable during all of these steps.			

Figure F-2. Targeting methodology.

DECIDE:

_______ Do the commander's planning guidance and intent contain enough detail to enable the targeting team to determine—
- — HVTs to nominate as HPTs?
- — Desired effects on each HPT?
- — When to attack each HPT?
- — Any restrictions or constraints?
- — Which HPTs require BDA?

_______ What targeting assets (organic, attached, supporting) are available to detect and attach HPTs?

_______ What detect, deliver, and assess support is needed from higher HQ?

_______ When must requests to higher HQ be submitted to obtain the support when required?

_______ Have target tracking responsibilities been established?

_______ Are systems in place to handoff the detected targets to assets that are capable of tracking them?

_______ What detect, deliver, and assess support is required from subordinate units; when is it required?

Figure F-3. Targeting checklist.

_______ What detect, deliver, and assess support requests have been received from subordinate units; what has been done with them?

_______ Has the AGM been synchronized with the DST and the maneuver and FS plans?

_______ Are all commands using a common datum for locations? If not, are procedures in place to correct differences in data?

DETECT:

_______ Does the collection plan focus on PIR HPTs? (Which includes HPTs designated as PIR?)

_______ What accuracy, timeliness, and validity standards (TSS) are in effect for detection and delivery systems?

_______ Are all TA systems fully employed?

_______ Have backup TA systems been identified for HPTs?

_______ Have responsibilities been assigned to the appropriate unit and/or agency for detecting each HPT?

_______ Are HPTs being tracked?

_______ Have verification procedures using backup systems been established where necessary?

_______ Are TA and BDA requirements distributed properly among systems that can accomplish both?

DELIVER:

_______ Have communications links been established between detection systems, the decisionmaker, and delivery systems?

_______ Have responsibilities been assigned to the appropriate unit and/or agency for attacking each HPT?

_______ Has a backup attack system been identified for each critical HPT? (The primary system may not be available at the time the HPT is verified.)

_______ Have FSCMs or AGMs and clearance procedures been established to facilitate target engagement?

Figure F-3. Targeting checklist (continued).

_______ Have O/O FSCMs or AGMs been established to facilitate future and transition operations?

_______ Have potential fratricide situations been identified; have procedures been established to positively control each situation?

_______ Have responsibilities been assigned to the appropriate unit or agency for tracking specific HPTs and providing BDA on specified HPTs?

_______ What are the procedures to update the HPTL and synchronize the AGM and DST if it becomes necessary to change the scheme of maneuver and FS as the situation changes?

ASSESS:

_______ Are the collection assets linked to specific HPTs still available?

_______ Have the collection asset managers been notified of the attack of a target requiring assessment?

_______ Have assessment asset managers been updated as to the actual target location?

_______ Has all the coordination for the assessment mission, particularly airborne assets, been accomplished?

_______ What is the status of BDA collection?

_______ Has the information from the mission been delivered to the appropriate agency for evaluation?

_______ Has the targeting team reviewed the results of the attack to determine restrike requirements?

_______ Has the target intelligence gathered from the assessment been incorporated into the overall enemy situation development?

Figure F-3. Targeting checklist (continued).

STEP 1 - DECIDE:

DECIDE provides the overall focus and sets priorities for intelligence collection and attack planning (Figure F-4). Targeting priorities must be addressed for each phase or critical event of an operation. The decisions made are reflected in visual products as follows:

- HPTL is a prioritized list of HPTs whose loss to the enemy will contribute to the success of the friendly COA.
- Intelligence collection plan, or R&S plan, answers the commander's PIR, to include those HPTs designated as PIR. The plan, within the availability of additional collection assets, supports the acquisition of more HPTs. Determining the intelligence requirements is the first step in the CM process. (See FMs 34-2 and 34-2-1.)
- TSS matrix addresses accuracy or other specific criteria that must be met before targets can be attacked.
- AGM, approved by the commander, addresses which targets will be attacked, how, when, and the desired effects.

The formats for the HPTLs, TSS, and AGMs are examples only. Targeting personnel must understand all that is involved in building these targeting tools. However, experienced staffs may prefer to develop their own tailored formats.

IPB and Mission Analysis. IPB, including EPB, is a continuous process and starts in the preparation phase. It continues to provide updated as well as new products and supports the targeting process. The intelligence estimate provides key support to target development. MCOOs, doctrinal, situation, event, and DSTs help identify ECOAs, NAIs, TAIs, HVTs.

TVA and Wargaming. The battlefield environment, its effects on combatants, and the threat are considered in IPB analysis to arrive at an intelligence estimate. TVA yields HVTs for a specific ECOA. Target spreadsheets identify the HVTs in relation to a type of operation, and target sheets (see FM 6-20-10, App A) give detailed targeting information for each HVT. This information is used during the IPB and the wargame processes. Both tools are developed by the G2 ASPS.

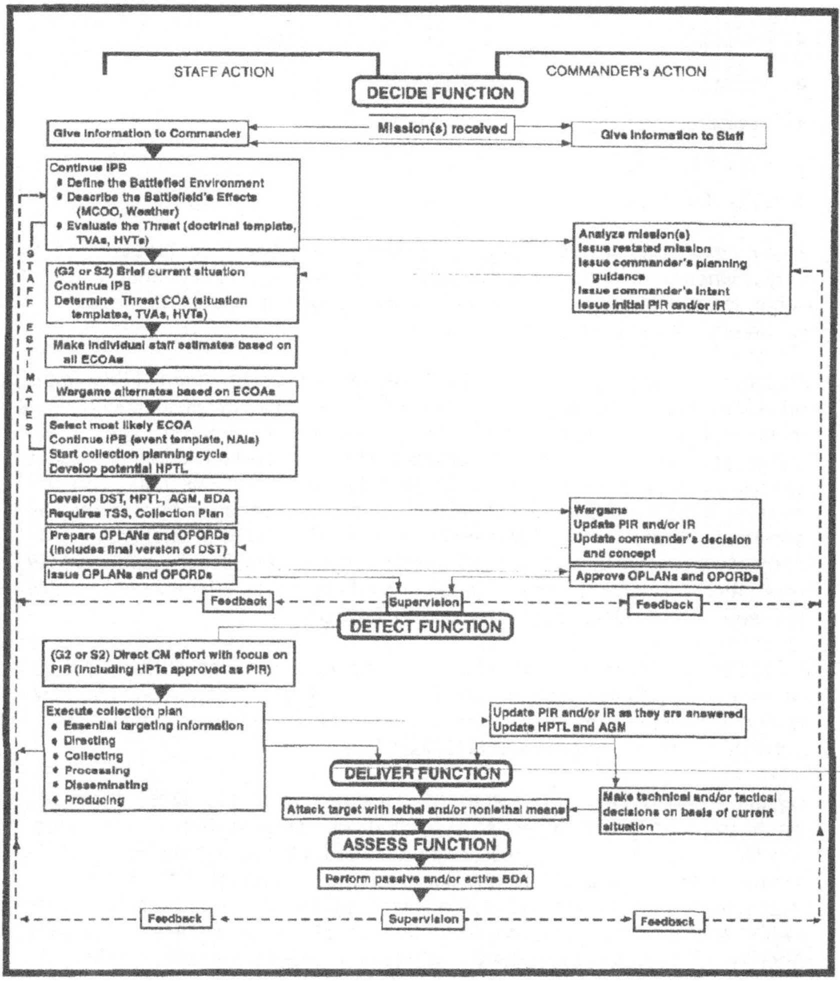

Figure F-4. The targeting process.

TVA, which is a detailed analysis of the enemy in selected COAs, uses the following enemy characteristics:

- Doctrine.
- Tactics.
- Equipment.
- Organizations.
- Expected behavior.

TVA methodology provides a relative ranking of target sets, or categories. It begins when target analysts in the G2/S2 assume the position of enemy commander. The target analyst, in coordination with other staff members, wargames the operation.

During wargaming, alternate friendly COAs are analyzed in terms of their impact on enemy operations and likely responses. The enemy battlefield functions that must be attacked to force the best enemy response are identified. The commander and staff analyze the criticality of friendly battlefield functions on a specific COA. The best places to attack HPTs are identified as TAIs. DPs or decision TPLs are used to ensure that the decision to engage or not to engage occurs at the proper time. Wargaming also identifies HVTs in priority that are critical to the success of the enemy mission. It also identifies the subset of HVTs which are HPTs that must be acquired and attacked for the friendly mission to succeed. HPTs selected are recorded on the DST.

Collection managers evaluate HVTs to determine asset detection capability. HPTs receive priority in the allocation of assets. The FSO determines friendly weapons systems capable of attacking HVTs with lethal fires, and the EWO nominates and coordinates to provide non-lethal fires.

Using the capabilities of the system to attack the targets, the G2/S2 analyzes and synthesizes the threat's response to each attack. Targets should be assigned priorities according to description, signature, degradation, and graphic representation. If targets have the same relative importance, a targeting team must prioritize the targets and seek advice from the FSE targeting analyst and the FAIO. After the targets are prioritized, they are placed on an HPTL (Figure F-5). After the HPTL is approved, the G2 will use it to develop the collection plan.

EVENT OR PHASE:		
PRIORITY	CATEGORY	TARGET

Figure F-5. Example high-payoff target list.

TSS. TSS are criteria applied to enemy activity (acquisitions and battlefield information) used in deciding whether the activity is a target. TSS break nominations into two categories: targets, which meet accuracy and timeliness requirements for attack; and suspected targets, which must be confirmed before any attack.

- HPT. This refers to the designated HPTs which the collection manager is tasked to acquire.
- TIMELINESS. Valid targets are reported to attack systems within the designated timeliness criteria.
- ACCURACY. Valid targets must be reported to the attack system meeting the required TLE criteria. The criteria is the least restrictive TLE considering the capabilities of available attack systems.

Considering these factors, different TSS may exist for a given enemy activity on the basis of different attack systems. For example, an enemy artillery battery may have a 150-meter TLE requirement for attack by cannon artillery and a 1 km requirement for attack helicopters. TSS are developed by the FSE in conjunction with MI personnel. Intelligence analysts use TSS to quickly determine targets from battlefield information and pass the targets to the FSE.

Attack system managers, such as FSEs, FCEs, or FDCs, use TSS to quickly identify targets for attack. Commands can develop *standard* TSS based on anticipated enemy OB and doctrine matched with the normally available attack systems.

The G2/S2 knows the accuracy of acquisition systems, associated TLE, and the expected dwell times of enemy targets. He can then specify whether information he reports to the attack system manager is a target or a suspected target. Some situations may require the system to identify friendly and neutral from threat before approval to fire is given. HPTs that meet all the criteria should be tracked until they are attacked IAW the AGM. Location of targets that do not meet TSS should be confirmed before they are attacked. The TSS can be depicted in a TSS matrix (Figure F-6).

The matrix lists each system that forwards targets directly to the FSE, FCE, or FDC. The effects of weather and terrain on the collection assets and on enemy equipment are considered. TSS are keyed to the situation. However, the greatest emphasis is on the enemy situation considering deception and the reliability of the source or agency that is reporting.

HPT	TIMELINESS	ACCURACY
COPs	3 hr	150 m
RISTA	30 min	150 m
2S3	30 min	500 m
M-46	30 min	500 m
ADA	15 min	500 m
CPs	3 hr	500 m
Ammunition	6 hr	1 km
Maneuver	1 hr	150 m

Figure F-6. Example target selection standards matrix.

Attack Guidance. Knowing target vulnerabilities and the effect an attack will have on enemy operations allows a staff to propose the most efficient available attack option. Key guidance is whether the commander wishes to disrupt, delay, limit damage, or destroy the enemy. During wargaming, DPs linked to events, areas (NAIs and TAIs), or points on the battlefield are developed. These DPs cue the command decisions and staff actions where tactical decisions are needed.

On the basis of commander's guidance, the targeting team recommends how each target should be engaged in terms of the effects of fire and attack options to use. Effects of fire (see Joint Pub 1-02) can be to **harass, suppress, neutralize,** or **destroy** the target. The subjective nature of what is meant by these terms means the commander must ensure the targeting team understands his use of them. Applying FS automation system default values further complicates this understanding.

- Harassing fire is designed to disturb the rest of the enemy troops, to curtail movement, and, by threat of losses, to lower morale. The decision to employ harassing fires needs careful consideration. Harassing fire has little real effect on the enemy, subjects gun crews to an additional workload, and increases the threat of counter-battery fires. ROE or the potential for adverse public opinion may prohibit its use. However, harassing fires may be a combat multiplier in some situations. Consider their use in SASO, delaying actions, and economy of force operations.

- Suppressive fire on or about a weapons system degrades its performance below the level needed to fulfill its mission objectives. Suppression lasts only as long as the fires continue. The duration of suppressive fires is either specified in the call for fire or established by SOP. Suppression is used to prevent effective fire on friendly forces. It is typically used to support a specified movement of forces. The FSCOORD needs to ask or calculate the *when* and *how long* questions.

- Neutralization fire is delivered to render the target ineffective or unusable temporarily. Neutralization fire results in enemy personnel or material becoming incapable of interfering with an operation or COA. Key questions the FSCOORD must ask are when and how long does the commander want the target to be neutralized. Most planned missions are neutralization fires.

- Destruction fire is delivered for the sole purpose of destroying material objects. It physically renders the target permanently combat-ineffective unless it is restored, reconstituted, or rebuilt. Setting automated FS default values for destruction of 30 percent does not guarantee achieving the commander's intent. The surviving 70 percent may still influence the operation. Destruction missions are expensive in terms of time and material. **Consider whether neutralization or suppression may be more efficient.**

The decision of what attack system to use is made at the same time as the decision on when to acquire and attack the target. Coordination is required when deciding to attack with two different means (such as EW and combat air operations). Coordination requirements are recorded during the wargame process.

The attack guidance must be approved by the commander and detail the following:

- A prioritized list of HPTs.
- When, how, and desired effects of attack.
- Any special instructions.
- HPTs that require BDA.

This information is developed during the wargame. Attack guidance—

- Applies to planned targets and targets of opportunity.
- May address specific or general target descriptions.
- Is provided to attack system managers via the AGM (Figure F-7).
- May change as the operation progresses.

The AGM must be updated during staff planning meetings and as the enemy situation changes. Consider separate AGMs for each phase of an operation.

PHASE/EVENT: Attack through the security zone				
HPTL	**WHEN**	**HOW**	**EFFECT**	**REMARKS**
COPs	P	GS ARTY	N	PLAN IN INITIAL PREP
RISTA and OPs	P	GS ARTY	N	PLAN IN INITIAL PREP
2S1 and 2S3	P	MLRS	N	PLAN IN INITIAL PREP
2S6, SA9 and SA13	P	GS ARTY	S	SEAD FOR AVN OPS
REGT CP	A	MLRS	N	
RESERVE BN	P	AVN BDE	D	INTENT TO ATTACK RESERVE BN IN EA HOT
LEGEND:	I = IMMEDIATE A = AS ACQUIRED P = PLANNED	S = SUPPRESS N = NEUTRALIZE D = DESTROY		
NOTES: [1]This is only an example of an AGM. Actual matrices are developed on the basis of the situation. [2]An "H" for harassing fires may be included in the EFFECT column during SASO.				

Figure F-7. Example attack guidance matrix.

HPTL. Lists the prioritized HPTs identified during wargaming. They have priority for engagement.

WHEN. Timing the attack of targets is critical to maximizing effects. During wargaming, the optimum time is identified and reflected in the WHEN column. A "**P**" indicates the target should not be engaged now but should be *planned* for future firing (e.g., a preparation, a SEAD program, or a countermobility program) or simply put on file. An "**A**" indicates such targets should be engaged in the sequence received in the headquarters, with respect to the priority noted in the HPTL. An "**I**" indicates the attack must be *immediate* and is a special case. This designation should be limited to a small percentage of targets and only for the most critical types. Too many immediate targets are disruptive and lower the efficiency of attack systems. Immediate attacks take precedence over all others and are conducted even if attack systems must be diverted from attacks already underway. Examples of important targets include:

- Missile systems capable of NBC attacks.
- Division headquarters.
- NBC weapons storage and support facilities.

MRLS may be considered for immediate attack depending on their demonstrated effectiveness against friendly forces and their tactical employment. The G3/S3 and FSCOORD or FSO must establish procedures within the TOC that allow for immediate attack of targets.

HOW. Links the attack system to the HPT. It is best to identify a primary and backup attack system for attack of HPTs.

EFFECT. Refers to the target attack criteria. The targeting team should specify attack criteria according to the commander's guidance. Target attack criteria should be quantifiable (e.g., percentage of casualties or destroyed elements, time, ordnance, and allocation or application of assets). Also, it can be noted as the number of battery or battalion volleys.

REMARKS. Notes which targets should not be attacked in certain tactical situations (e.g., if the enemy is withdrawing). Examples of how to use this column:

- Accuracy or time constraints.
- Required coordination.
- Limitations on the amount or type of ammunition.
- Any need for BDA.

STEP 2 - DETECT:

DETECT is the next critical function in the targeting process. The G2/S2 is the main figure in directing the effort to detect identified HPTs. To identify who, what, when, and how for TA, the G2/S2 must work closely with the—

- ACE.
- FAIO.
- Targeting officer and FSO.

This process determines accurate, identifiable, and timely requirements for collection systems. The ACE must ensure collection system asset managers understand these requirements.

Information needs for target detection are expressed as PIR and IR. Their priority depends on the importance of the target to the friendly COA and tracking requirements. PIR and IR that support detection of HPTs are incorporated into the overall unit collection plan.

Targets are detected by the maximum use of all available assets. The G2/S2 must focus the intelligence acquisition efforts on the designated HPTs and PIR. Situation development information, through detection and tracking, will be accumulated as collection systems satisfy PIR and IR. The collection manager—

- Considers availability and capabilities of all collection assets within his echelon and those available to subordinate, higher, and adjacent units.
- Considers joint or combined force assets.
- Translates the PIR and IR into SIR and SOR.
- Arranges direct dissemination of targeting information from the collector to the targeting cell or targeting intelligence to the FSE if possible.

Detection Procedures. Use all TA assets effectively and efficiently. Avoid duplication of effort among intelligence available assets unless it is required to confirm target information. At corps and division, the ACE develops and manages the collection plan and ensures that no gaps in planned collection exist. This allows timely combat information to be collected to answer the commander's intelligence and TA requirements. To detect HPTs, give clear and concise taskings to those TA systems capable of detecting a giventarget. This information lets analysts develop the enemy situation and identify targets (Figure F-8).

FS personnel provide the G2/S2 with the degree of located accuracy. The accuracy requirements are matched to the TLE of the collection systems. This allows the G2/S2 to develop more detailed TSS. NAIs and TAIs identified are matched with the most capable detection system available. If the target type and its associated signatures (e.g., electronic, visual, thermal) are known, the most capable collection asset can be directed against the target. The asset can be placed in the best position according to estimates of when and where the enemy target will be located.

Assets are positioned based on friendly estimates of where the enemy target is located. Information needed to detect targets are expressed in PIR and IR, which are incorporated into the collection plan. The collection manager translates the PIR and IR into SIR.

The collection manager considers the availability of all collection assets at all echelons. When a target is detected, the information is passed to the FAIO to determine if it is an HPT, its priority, and if it meets the TSS. To ensure that target information is passed quickly, the FAIO should be located in the ACE with communications to the FSE. Once an HPT is discovered, the FAIOs coordinate with their respective G2s/S2s and pass the targets directly to the FSE. If the commander approves the target, it is then passed to a firing unit. Mobile HPTs must be detected and tracked to maintain a current target location. Target tracking is inherent to detection.

BOS	TARGET ACQUISITION MEANS
INTELLIGENCE	SIGINT/EW, IMINT, HUMINT, TECHINT, MASINT systems.
MANEUVER	Patrols, scouts, units in contact, air/ground cavalry/air units.
FS	TAB, AN/TPQ-36/37 radars, FOs, AFSOs, COLTs.
AD	ADA scouts, ADA system's acquisition radars.
MOBILITY, COUNTER-MOBILITY, SURVIVABILITY	Engineer and NBC reconnaissance.
CSS	Truck drivers, base cluster recon/counterrecon patrols.
C^2	MIJI reports.
AIR FORCE	Joint STARS, AWACS.

Figure F-8. BOS detection capabilities.

As the assets collect information for target development, the information is forwarded to ACE intelligence analysts who use the information to perform situation and target development. Once a target specified for attack is identified, the analysts pass it to the FSE. The FSE executes the attack guidance against the target. Coordination between the intelligence staff and the FSE is essential to ensure that the targets are passed to an attack system that will engage the target.

The FAIOs can coordinate with the G2 and FSE to pass HPTs and other targets directly to the FCE at the division artillery TOC or, if approved by the maneuver commander, directly to a firing unit. The result is an efficient attack of targets which have been designated in advance for attack.

When the FAIO gets intelligence information which warrants attack, he notifies the FSE. This allows the FAIO to focus on intelligence analysis and the FSE to manage the control of fires. The targeting officer at the maneuver brigade and the S2 at the battalion perform FAIO functions.

Tracking priorities are based on the commander's concept of the operation and targeting priorities. Tracking is executed through the collection plan. Not all targets will be tracked. Because many critical targets move frequently or constantly, these HPTs require tracking.

Synchronization of Intelligence Collection. While conducting operations, the G2/S2 monitors execution of the collection plan. He uses an ISM to synchronize the intelligence system in order to give answers to PIR in time to affect the decisions they support. As HPTs are identified, he informs the appropriate FSE and cross-cues collection assets to support BDA as needed.

During delivery, NRT collectors and observers are cued to continue tracking targets during their engagement. Planned or cued BDA collection and reporting help determine if the engagement produced the desired effects. If not, continued tracking supports immediate reengagement.

To facilitate hand-off of target and tracking responsibilities, the G2/S2 coordinates with higher and subordinate units. Requirements for intelligence in a given area can then be directed to the unit responsible for that area. Operational graphics are useful when tracking particular threat units or HPTs. Passing responsibility for the target as it crosses the graphic ensures that it is not lost in crossing.

Synchronization continues during the CM process. The G2/S2 uses the products of the targeting process to develop SIR and SOR that synchronize the collection effort with command DPs. He plans and allocates time for collection, processing, and dissemination.

During execution the collection manager monitors the collection and reporting to ensure synchronization. He prompts asset managers to meet timelines, cross-cueing, and retasking as necessary to keep intelligence operations synchronized with command operations.

IEW Systems. The collection manager at each echelon gives appropriate IEW agencies missions through the collection plan and ISM. The intelligence analyst directly supports the collection manager. To properly synthesize information being received from forward units and collection systems intelligence analysts must understand their capabilities and limitations.

Figure F-9 explains the intelligence disciplines that provide targeting information through collection.

DISCIPLINE	FUNCTION
SIGINT/EW	• Friendly SIGINT/EW systems search for, intercept, and locate threat electronic signals emanating from radios, radars, and beacons. • Signatures are caused by electronics associated with the particular C^2 functions of the target. • Signatures identify signals based on characteristics of the signal or content of any messages intercepted. • The signals are then associated with an adversary's OB. • Electronic systems organic to tactical and operational (corps and theater) EW units can locate threat radio, radar, beacon, and C^2 system emissions using DF equipment. • DF efforts have inherent considerations that affect target accuracy such as DF system accuracy, terrain, and distance to target.
IMINT	• Imagery assets can be located in the AEB of the corps MI Bde, GRCS, and GS and DS companies of the Div MI Bn TUAV when fielded. • TUAV is organic to the DS Co of a heavy division. • TUAV is organic to the GS Co in a light infantry division. • FCE is assigned to the GS Co. • Predator UAV and Joint-STARS are USAF assets. They include airborne and ground-based IMINT sensors such as photo, TV, infrared, and MTI radar. • Targets have visual and infrared signatures that can be imaged. • Imagery analysts study results of IMINT missions to determine the presence, location, and assessed damage to threat targets. These films and devices can detect thermal signatures produced by equipment (engines, heaters, or any device producing or absorbing heat). These devices can also differentiate camouflaged equipment and structures. • Mounted sensors in UAVs provide TA capability which can locate threat elements unseen by ground observers and

Figure F-9. Intelligence disciplines.

	can successfully direct engagement of the targets. SLAR can detect the movement of vehicles beyond visual range of ground observers. • Imagery and intelligence analysts look for the threat's patterns of activity or intentions produced from IMINT systems. • Patterns of movement (numbers, direction, times) are analyzed and synthesized with other combat information to determine a threat commander's intentions. • Imagery can be requested from national and theater systems to support area studies and operations. • Imagery can also be used for training in garrison and at CTCs.
HUMINT	• HUMINT is produced by LRSUs, CI teams that are positioned throughout the depth of the battlefield, and interrogation teams that may be at maneuver battalions and higher. • Interrogation of threat prisoners can be focused on the detection of HVT and HPTs. • Interrogators should receive lists of intelligence interests and the commander's PIR. • Giving current HPTLs to the interrogators allows them to focus their efforts on locating these targets while gleaning additional information for situation development. • LRSUs are another HUMINT source. They may be tasked to determine the presence or absence of threat activity within the NAIs developed during the IPB process. • Keying these NAIs to HPTs allows the LRSUs to contribute significantly to the target detection process. • Information gained from LRSUs is real or NRT and can result in immediate engagement. • Due to extensive preparation and support requirements, LRSUs should be planned early and used IAW its doctrine (e.g., depth, target selection, insertion techniques). • CI teams detect subversion, espionage, and terrorist threats to friendly forces, and can also work with the local population to gather items of intelligence interest. • SOF may be tasked through SOCCE assigned at brigade, division, or Corps.

Figure F-9. Intelligence disciplines (continued).

TECHINT	• The TECHINT system is made up of two parts: S&TI and battlefield TECHINT. • S&TI supports strategic level of war intelligence. • Battlefield TECHINT supports operational and tactical levels of war intelligence. • TECHINT aids the commander by providing products that either identify or counter an adversary's monetary technological advantage. • TECHINT is an integral part of all-source intelligence because it involves everyone from individual soldiers to the policy makers. • Foreign material is photographed or captured and exploited for information, capabilities, limitations, and vulnerabilities. • The 203d MI Battalion operates the CMEC for the Army, or forms the Army's component of JCMEC or CCMEC. • Teams are sent from the battalion down to ECB units. • The G2, G3, TECHINT commander, and MI commander comprise the battlefield TECHINT team. • The commanders execute the staff's directives to collect materials for the TECHINT teams to exploit.
MASINT	• A MASINT sensor collects energy emitted by or reflected from an object or event for the purpose of identification and characterization. • Enemy weapon systems have physical characteristics and distinct performance identifiers when used. • Dynamic targets and events are primary candidates for MASINT exploitation by the MI force projection brigades. • These sources create energy emissions across the EM spectrum which carry information about its characteristics and performance via waves or particles. • This information is the signature collected by MASINT sensors; it can be used when developing threat assessments, OBs, IPB, and targeting materials. • MASINT data becomes invaluable during the rapid reprogramming of *smart* weapons and sensors or when changing tactics.

Figure F-9. Intelligence disciplines (continued).

STEP 3 - DELIVER:

DELIVER executes the target attack guidance and supports the commander's battle plan once the HPTs have been located and identified (Figure F-10).

ATTACK OF TARGETS MUST:	■ Satisfy attack guidance developed in *decide* function. ■ Require two categories of decisions—*tactical* and *technical*.
TACTICAL DECISIONS DETERMINE:	■ Desired effect, degree of damage, or both. ■ Attack system to be used. ■ Time of attack. This is determined IAW type of target: Planned target or target of opportunity. **PLANNED TARGET:** –Some targets will not appear as anticipated. –Target attack takes place only when the forecasted enemy activity occurs in projected time or place. Detection & tracking of activities associated with target becomes trigger for target attack.
TARGETING TEAM:	
G2 RESPONSIBILITIES:	■ Verifies enemy activity as the planned target to be attacked. ■ Validates target by conducting final check of reliability of source & accuracy (time & location) of target. Passes target to FSE. ■ Current operations officer checks legality of target in terms of the ROE.
FSE RESPONSIBILITIES:	■ Determines if attack systems planned are available & still best systems for attack. ■ Coordinates with higher/lower/adjacent units, other services, allies & HN (important where potential fratricide situations are identified).

Figure F-10. Deliver functions and responsibilities.

	■ Issues fire mission request to appropriate executing units. ■ Informs G2 of target attack. G2 alerts appropriate system responsible for BDA (when applicable). **TARGETS OF OPPORTUNITY:** –Are processed the same as planned HPTs. –Those not on HPTL are first evaluated to determine when or if they should be attacked. ■ Decision to attack follows attack guidance & is based on— •Activity of the target. •Dwell time. •Target payoff compared to other targets currently being processed for engagement. –If decision is made to attack immediately, target is processed further. –Availability & capabilities of attack system to engage targets are assessed. –If target exceeds availability or capabilities, send target to higher HQ for immediate attack. –If decision is to defer attack, continue tracking; determine DPs for attack, and modify collection taskings as appropriate.
TECHNICAL DECISIONS (DETERMINED BASED ON TACTICAL DECISIONS):	■ Precise delivery means. ■ Number and type of munitions. ■ Unit to conduct the attack. ■ Response time of the attacking unit. ■ Results in physical attack of targets by lethal and/or nonlethal means. ■ FS cell directs attack system to attack target once tactical decisions have been made. ■ FS cell provides attack system manager with—

Figure F-10. Deliver functions and responsibilities (continued).

	•Selected time of attack. •Effects desired IAW previous discussion. •Any special restraints or requests for particular munitions types. ■ Attack system manager (FSCOORD, ALO, AVN BDE, LNO, NGLO) determines if his system can meet requirements. If cannot, he notifies FS cell. Some reasons for not being able to meet requirements include— •System not available at specified time. •Required munitions not available. •Target out of range. ■ FS cell must decide if selected system should attack under different criteria or if a different system should be used.
TARGETS OF OPPORTUNITY ARE ATTACKED BASED ON:	■ The target's activity. ■ Estimated assembly area activity.
DESIRED EFFECTS: **DISRUPT** **DELAY** **LIMIT**	■ Planned Targets: •Verify enemy activity as that planned to be attacked. •Reaffirm decision to attack. •Issue the fire mission request (through FS cell) to appropriate executing units. ■ Targets of Opportunity: •Targeting team must decide payoff and availability of attack systems and munitions.
ATTACK SYSTEM:	■ Planned Targets: •Decision made during *decide* function. •Determine system availability & capability. •Targeting team must determine best system available to attack target if system not available or capable.

Figure F-10. Deliver functions and responsibilities (continued).

	■ Targets of Opportunity: •Targeting team must determine attack system, subject to maneuver commander's approval. •Consider all available attack systems. •Attacking targets should optimize capabilities of— –Light and heavy ground forces. –Attack helicopters. –FA. –Mortars. –Naval gun fire. –Combat air operations (CAS and air interdiction). –Offensive EW. •Consider availability & capabilities of each resource using the following: –Desired effects on the target. –Payoff of the target. –Degree of risk to use asset against target. –Impact on friendly operations. •Target attack must be coordinated among two or more attack systems. •Engaging a target by lethal means, along with jamming or monitoring, may be of greater benefit than simply firing at the target.

Figure F-10. Deliver functions and responsibilities (continued).

STEP 4 - ASSESS:

Combat assessment is the determination of the effectiveness of force employment during military operations. It is composed of three elements:

- BDA, which is the timely and accurate estimate of damage resulting from the application of lethal or nonlethal military force against a target.

- MEA, which is an assessment of the military force in terms of weapon systems and munitions effectiveness.

- Reattack recommendation.

In combination, BDA and MEA inform the commander of effects against targets and target sets. Based on this, the threat's ability to make and sustain war and centers of gravity are continuously estimated. During the review of the effects of the campaign, restrike recommendations are proposed or executed. BDA pertains to the results of attacks on targets designated by the commander. Producing BDA is primarily an intelligence responsibility, but requires coordination with operational elements. BDA requirements must be translated into PIR. It accomplishes the following:

- At the tactical level, commanders use BDA to get a series of timely and accurate snapshots of their effect on the enemy. It provides commanders an estimate of the enemy's combat effectiveness, capabilities, and intentions. This helps commanders determine when or if their targeting effort is accomplishing their objectives.

- BDA helps to determine if restrike is necessary. Commanders use BDA to allocate or redirect attack systems to make the best use of available combat power.

The G3 through the targeting team conducts MEA concurrently and interactively with BDA as a function of combat assessment. MEA is used as the basis for recommending changes to increase effectiveness in—

- Methodology.
- Tactics.
- Weapon systems.
- Munitions.
- Weapon delivery parameters.

The G3 develops MEA by determining the effectiveness of tactics, weapons systems, and munitions. Munitions effect on targets can be calculated by obtaining rounds fired on specific targets by artillery assets. The targeting team may generate modified commander's guidance concerning—

- UBL.
- RSR.
- CSR.

The need for BDA for specific HPTs is determined during the **decide** function. Record BDA on the AGM and intelligence collection plan. The resources used for BDA are the same resources used for target development and TA. An asset used for BDA may not be available for target development and TA. The ACE receives, processes, and disseminates to the targeting team the results of attack, which are analyzed in terms of desired effects.

The targeting team must keep the following BDA principles in mind:

- BDA must measure things that are important to commanders, not make important the things that are easily measurable.
- BDA must be objective. When a G2/S2 receives a BDA product from another echelon, he should verify the conclusions (time permitting). G2s/S2s at all echelons must strive to identify and resolve discrepancies among the BDA analysts at different headquarters.
- The degree of reliability and credibility of the assessment relies largely upon collection resources. The quantity and quality of collection assets influence whether the assessment is highly reliable (concrete, quantifiable, and precise) or has low reliability (best guess). The best BDAs use more than one collection discipline to verify each conclusion.

Each BDA has three assessment components (Figure F-11). Each requires different sensors, analytical elements, and timelines. They are not necessarily subcomponents of each BDA report. (See FM 6-20-10, Appendix D.)

COMPONENTS	DESCRIPTION
PDA	• Qualitative physical damage from munitions blast, fragmentation, and fire. • Based on observed or interpreted damage.

FDA	• Effects of military force to degrade or destroy capability of target to perform mission. • Assessment based on all-source intelligence. • Includes an estimate of time to reconstitute or replace target. • Temporary assessment compared to a TSA used for specific missions.
TSA	• The overall impact of the full spectrum of military operations on an entire target system's capability. • Can be applied against an adversary's combat effectiveness. • May address significant subdivisions of a target. • A more permanent assessment.

Figure F-11. BDA functions.

BDA is more than determining the number of casualties or the amount of equipment destroyed. The targeting team can use other information, such as—

- Whether the targets are moving or hardening in response to the attack.
- Changes in deception efforts and techniques.
- Increased communication efforts as the result of jamming.
- Whether the damage achieved is affecting the enemy's combat effectiveness as expected.

BDA may also be passive by compiling information regarding a particular target or area (e.g., the cessation of fires from an area). If BDA is to be made, the targeting team must give intelligence acquisition systems adequate warning for sensors to be directed at the target at the proper time. BDA results may change plans and earlier decisions. The targeting team must periodically update the decisions made during the **decide** function concerning—

- IPB products
- HPTLs.
- TSS.
- AGMs.
- Intelligence collection plans.
- OPLANs.

Based on BDA and MEA, the G2/G3 consider the level to which operational objectives have been achieved and make recommendations to the commander. Reattack and other recommendations should address operational objectives relative to the—

- Target.
- Target critical elements.
- Target systems.
- Enemy combat force strengths.

BDA key players are the commander, operations officer, FSO, AAO, ALO, and S2. The S2 integrates intelligence and operational data. In coordination with the G3, the G2/S2—

- Recommends HPTs.

- Develops and recommends IR to include those for targeting and BDA. Some requirements become PIR.

- Coordinates with the G3/S3, AAO, and FSO to develop a fully coordinated targeting and BDA plan.

- Develops a synchronized collection plan to answer the commander's IR and tracks and maintains BDA charts and files.

- Tasks or requests collection support from the appropriate unit or agency to collect information required to satisfy the commander's targeting objectives and BDA reporting requirements.

- Establishes procedures to ensure reports from FOs, scouts, troops in contact, pilots, and AVTRs are quickly made available for BDA analysis.

- Matches BDA reporting requirements against the commander's objectives to determine targeting effort drain; develops and maintains historical BDA databases, and disseminates hard and soft copy intelligence and BDA results.

- Uses the results of BDA and combat assessment to determine the need to develop further threat COAs.

  — Determines priority for collection assets between the targeting effort and the BDA supporting requirements.

— Determines and updates enemy capabilities based upon results of the targeting effort.

SASO CONSIDERATIONS

Target development and targeting are difficult in an SASO environment. Although the principles of the targeting process still apply in SASO, consider the following:

- ROE tend to be more restrictive than in conventional combat situations. Everyone involved in the targeting process must completely understand the ROE.

- Identification of HVTs, threat intentions, threat locations, and weapons systems is slow to develop and may not be able to be acted upon when fully developed.

- Because of the nature of the threat (usually a relatively small, unconventional enemy may constitute a significant threat to US forces) determination of HVTs (specifically, what constitutes an HVT) is extremely difficult. Thus, it is difficult to develop precise time, location, and rates of march for a threat force.

- "Targets," which blend into the population, are often less visible than conventional targets.

- When identified, targets present less of a window of opportunity in which to react, decreasing decision time.

- Because of considerations of the HN population, targeting must be precise. Selecting the right weapons system for the right target is paramount.

- Interaction and coordination with nontraditional elements and agencies may be required. This may include private organizations, NGOs, other federal agencies, civil agencies, allies or coalition forces, and HN forces.

CARVER TECHNIQUE

ARSOF uses CARVER factors to assess mission, validity, and requirements. They are also used in technical appreciation and target analysis.

The CARVER selection factors assist in selecting the best targets or components to attack. As the factors are considered, they are given a numerical value, which represents the desirability of attacking the target. The values are then placed in a decision matrix. After CARVER values for each target or component are assigned, the sum of the values indicates the highest value target or component to be attacked within the limits of the statement of requirements and commander's intent. Figure F-12 explains the CARVER techniques.

	DESCRIPTION	FACTORS	CRITERIA & VALUES*	SCALE
Criticality	•Means target value. •Primary consideration in targeting. •Target is critical when its destruction or damage has major impact on military, political, or economic operations. •Consider targets within a system in relation to other elements of the target system. •Value of target will change as situation develops. •Target requires use of time-sensitive methods which respond to changing situation. **EXAMPLE:** When one has few locomotives, railroad bridges may be less critical as targets; however, safe-guarding bridges may be critical to maneuvering conventional forces requiring use of such bridges.	•Time—How rapidly will impact of target attack affect operations? •Quality—What % of output, production, or service will be curtailed by target damage? •Surrogates—What will be effect on output, production, and service? •Relativity—How many targets? What are their positions? How is their relative value determined? What will be affected in the system or complex stream?	•Immediate halt; target cannot function without it. •Halt within 1 day, or 66%. •Halt within 1 week, or 33%. •Halt within 10 days, or 10%. •No significant affect. *Refers to time or %; curtailment in output, production, or service.	9 to 10 7 to 8 5 to 6 3 to 4 1 to 2

Figure F-12. CARVER techniques.

	DESCRIPTION	FACTORS	CRITERIA & VALUES*	SCALE
Accessibility	•Target is accessible when operations element can reach the target with sufficient personnel and equipment to accomplish mission. •Entails identifying & studying critical paths the operational element must take to achieve objectives & measuring those things that aid or impede access. •You must be able to reach target AND remain there for extended periods. •Basic steps identifying accessibility: –Infiltration from staging base to target. –Movement from point of entry to target/objective. –Movement to target's critical element. –Exfiltration.	•Active & passive EWS. •Detection devices. •AD capabilities within target area. •Road & rail transportation systems. •Type terrain & its use. •Concealment & cover. •Population density. •Other natural or synthetic obstacles & barriers. •Current & climatic weather conditions. •Analysis along each critical path to target should measure time it would take for action element to bypass, neutralize, or penetrate barriers/ obstacles. •Measured by relative ease or difficulty of movement for operational element & likelihood of detection. Consider using standoff weapons in such evaluations.	•Easily accessible standoff weapons can be employed. •Inside a perimeter fence but outdoors. •Inside a building but on ground floor. •Inside a building but on second floor or in a basement; climbing or lowering is required. •Not accessible or inaccessible without extreme difficulty.	9 to 10 7 to 8 5 to 6 3 to 4 1 to 2

Figure F-12. CARVER techniques (continued).

	DESCRIPTION	FACTORS	CRITERIA & VALUES*	SCALE
Recuper-ability	•Measured in time—How long will it take to replace, repair, or bypass destruction of or damage to the target? •Varies with sources and type of targeted components and availability of spare parts availability.	•Onhand equipment such as railroad cranes, dry docks, and cannibalization. •Restoration and substitution through redundancies. •Onhand spares. •Equivalent OB equipment sets that backup critical equipment or components; effects of economic embargoes and labor unrest.	Replacement, repair, or substitution requires—	
			•1 month or more.	9 to 10
			•1 week to 1 month.	7 to 8
			•72 hours to 1 week.	5 to 6
			•24 to 72 hours.	3 to 4
			•Same day replacement, repair, or substitution.	1 to 2
Vulnerability	•Target is vulnerable if operational element has means & expertise to successfully attack the target. •Scale of critical component needs to be compared with capability of attacking element to destroy or damage it. •Attacking element may tend to— –Choose special components. –Do permanent damage.	•Depends on— –Nature and construction of the target. –Amount of damage required. –Assets available, (e.g., personnel, expertise, motivation, weapons, explosives, and equipment.)	•Vulnerable to long-range laser target designation, small arms fire, or charges of 5 lbs or less.	9 to 10
			•Vulnerable to light antiarmor weapons fire or charges of 5 to 10 lbs.	7 to 8

Figure F-12. CARVER techniques (continued).

	DESCRIPTION	FACTORS	CRITERIA & VALUES*	SCALE
	–Maximize effects through use of onsite materials. –Cause the target to self-destruct.		•Vulnerable to medium antiarmor weapons fire, bulk charges of 10 to 30 lbs, or very careful placement of smaller charges.	5 to 6
			•Vulnerable to heavy antiarmor fire, bulk charges of 30 to 50 lbs, or requires special weapons.	3 to 4
			•Invulnerable to all but the most extreme targetting measures.	1 to 2
Effect	•Measure of possible military, political, economic, psychological & sociological impacts at target & beyond. •Closely related to the measure of target criticality. •Type & magnitude of given effects desired will help planners select targets & target components.	•Primary effect of destruction of 2 adjacent long-range radar sites in EWS may be to open hole in system that is of sufficient size & duration to permit attacker to launch a successful air or missile nuclear strike against defender.	•Overwhelmingly positive effects; no significant negative effects.	9 to 10
			•Moderately positive effects; few significant negative effects.	7 to 8
			•No significant effects; neutral.	5 to 6
			•Moderately negative effects; few significant positive effects.	3 to 4

Figure F-12. CARVER techniques (continued).

	DESCRIPTION	FACTORS	CRITERIA & VALUES*	SCALE
	•Addresses all significant effects, whether desired or not, that may result once selected target component is attacked. •Traditionally, this element addressed effect on local population; now there are broader considerations. •Frequently neutral at tactical level.	•Can include— –Triggering of counter-measures. –Support of negation of PSYOP themes. –Unemploy-ment. –Reprisals against civilian popu-lace. –Collateral damage to other targets. •Possible effects can be specula-tive & thus should be labeled. •Effects of same attack may be different at tactical, opera-tional & strategic levels. **EXAMPLE:** Destruction of a substation may not affect local power but cuts off power to adjacent region.	•Overwhelmingly negative effects; no significant positive effects.	1 to 2

Figure F-12. CARVER techniques (continued).

	DESCRIPTION	FACTORS	CRITERIA & VALUES*	SCALE
Recognizability	•Target's recognizability is degree to which it can be recognized by an operational element or intel collection & recon assets under varying conditions. •Weather impacts visibility. •Rain, snow & ground fog may obscure observation. •Road segments with sparse vegetation & adjacent high ground provide excellent conditions for good observation. •Distance, light & season must also be considered.	•Site & complexity of target. •Existence of distinctive target signatures. •Presence of masking or camouflage. •Technical sophistication & training of attackers.	•Target is clearly recognizable under all conditions & from a distance; requires little or no training.	9 to 10
			•Target is easily recognizable at small-arms range & requires small amount of training.	7 to 8
			•Target is difficult to recognize at night or in bad weather, or might be confused with other targets or components; requires some training.	5 to 6
			•Target is difficult to recognize at night or in bad weather, even within small-arms range; it is easily confused with other targets or components; requires extensive training.	3 to 4
			•Target cannot be recognized under any conditions, except by experts.	1 to 2

	DESCRIPTION	FACTORS	CRITERIA & VALUES*	SCALE

Figure F-12. CARVER techniques (continued).

CARVER factors and their assigned values are used to construct a CARVER matrix. Table F-1 is a tool for rating the desirability of potential targets and allocating attack resources.

- List the potential targets in the left column.
- For strategic level analysis, list the enemy's systems or subsystems (electric power supply, rail system).
- For tactical level analysis, list the complexes or components of the subsystems or complexes selected for attack by your higher HQ.
- As each potential target is evaluated for each CARVER factor, enter the appropriate value into the matrix.
- Once all the potential targets have been evaluated, add the values for each potential target.
- The sums represent the relative desirability of each potential target; this constitutes a prioritized list of targets.
- Attack those targets with the highest totals first.
- If additional men or munitions are available, allocate these resources to the remaining potential targets in descending numerical order.
- This allocation scheme will maximize the use of limited resources.

The S2 can use the CARVER matrix to present operation planners with a variety of attack options. With the matrix, he can discuss the strengths and weaknesses of each COA against the target. Having arrived at conclusions through the rigorous evaluation process, the S2 can comfortably defend his choices.

Table F-1. Bulk electric power supply.

POTENTIAL TARGETS	C	A	R	V	E	R	TOTAL
Fuel Tanks	8	9	3	8	5	6	41
Fuel Pumps	8	6	2	10	5	3	34
Boilers	6	2	10	4	5	4	31
Turbines	8	6	10	7	5	9	45
Generators	4	6	10	7	5	9	41
Condenser	8	8	5	2	5	4	34
Feed Pumps	3	8	5	8	5	6	33
Water Pumps	3	8	5	8	5	4	33
Generator Step up Transformer	10	10	10	9	5	9	53

APPENDIX G

IEW SYSTEMS

QUICKFIX (AN/ALQ-151(V1)) Advanced QUICKFIX (AN/ALQ-151(V2))

FUNCTION: VHF DF intercept, VHF EA, VHF DF (nets w/TRAILBLAZER for DF). Component of IEWCS AF/VHF/UHF ES

PRIME MOVER: EH-60A, EH-60L (BLACKHAWK)

UNIT AND QTY: HVY DIV: 3 Systems 3 Flt Plt, Avn Bde; LT DIV: 3 Systems, 3 Flt Plt, Avn Bde; AASLT DIV: 3 Systems, 3 Flt Plt, HHOC, MI Bn; ABN DIV: 3 Systems, 3 Flt Plt, Cbt Avn Sqdn; ACR: 3 Systems, 3 Flt Plt, Cbt Avn Sqdn

INTEL APPLICATION: COMINT

Airborne Reconnaissance Low (ARL)

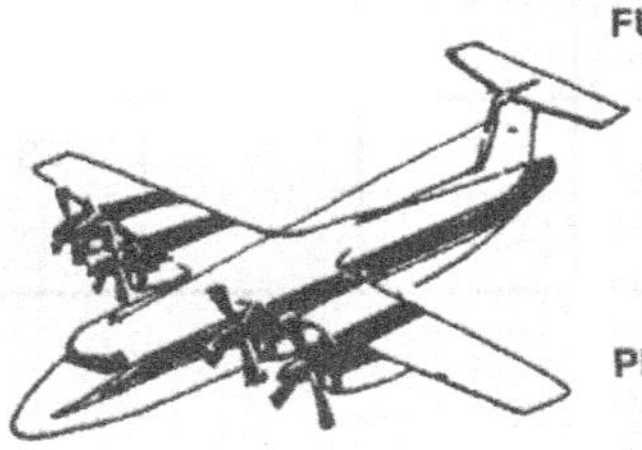

FUNCTION: Ground processing station, ACT 101 remote rcv sets, TACLINK video rcv sets, FLIR system, daylight imaging systems, infrared line scanner, radio intercept/DF systems; interoperates w/ASAS & CTT

PRIME MOVER: DHC-7 arrives ready for immediate employment

UNIT AND QTY: Select Brigades

INTEL APPLICATION: COMINT/IMINT

Commander's Tactical Terminal (CTT), AN/TSC-125

FUNCTION: Communications terminal; GUARDRAIL downlink on TRIXS network; RRS on board RC-12 & U2-R, RRTS; includes OT, RBP, RRT, LOS adaptive array antenna, SDS, embedded COMSEC

PRIME MOVER: NA

UNIT AND QTV: 1- MI at all echelons

INTEL APPLICATION: SIGINT

Joint Tactical Terminal: JTT/H3 (AN/USC-55)
JTT/H-R3 (AN/USR-6)

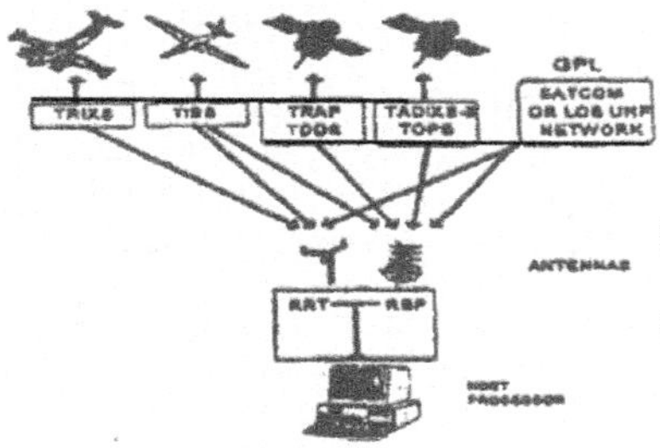

FUNCTION: Communications terminal, 2 LRUs, 1 RBP, 1 RRT, host processor & antenna array (both user provided); mobile & mountable in F/W & R/W, tracked & wheeled vehicles, shelters

PRIME MOVER: NA

UNIT AND QTY: Avn, ADA, FA, MI at all echelons; integrated into GSM/CGS, ASAS, ETRAC, THAADS, Patriot, JTAGS, TROJAN SPIRIT, AFATDDS, other multiservice systems

INTEL APPLICATION: SIGINT

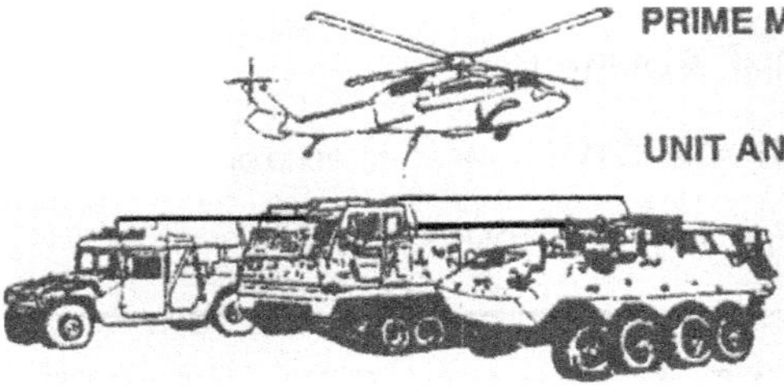

Ground-Based Common Sensor (GBCS)

FUNCTION: ES (intercept& DF); EA (jamming); COMINT targeting location data; common subsystems with AQF & MEWSS

PRIME MOVER: EFVS (HVY DIV) or HVY HMMWV (LT DIV)

UNIT AND QTY: 6 GBCS-H per MI Bn (H); 4 GBCS-H per MI Co (ACR); 4 GBCS-L per MI Bn (ABN, AASLT, LT); 6 per MI Bn (H, LT, ABN, AASLT, DIV); 6 per MI Co; ACR

INTEL APPLICATION: COMINT/ELINT

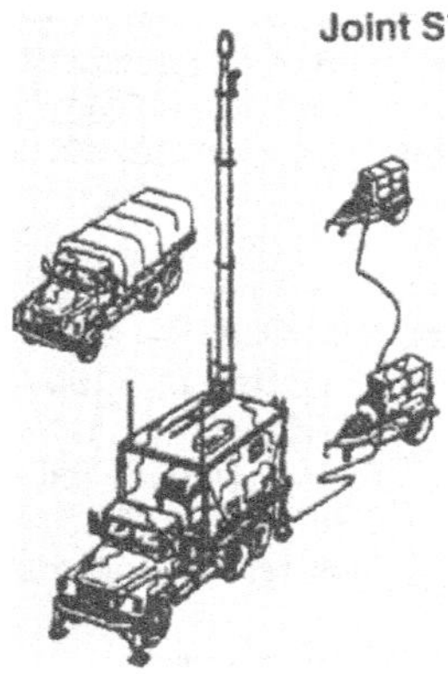

Joint STARS Ground Station Module (GSM) (AN/TSQ-132)

FUNCTION: MTI, FTI, SAR imagery; supports situation development, battle mgt, targeting, force protection, limited BDA/TDA, operations planning

PRIME MOVER: LT - 2xHMMWV; MED - 2X5-ton HMMWV

UNIT AND QTY: 6- Selected DIV; 6- Selected Corps; 1- ACR; 1- EAC

INTEL APPLICATION: IMINT/SIGINT/INTEL/ ARTY/AVN SUPPORT

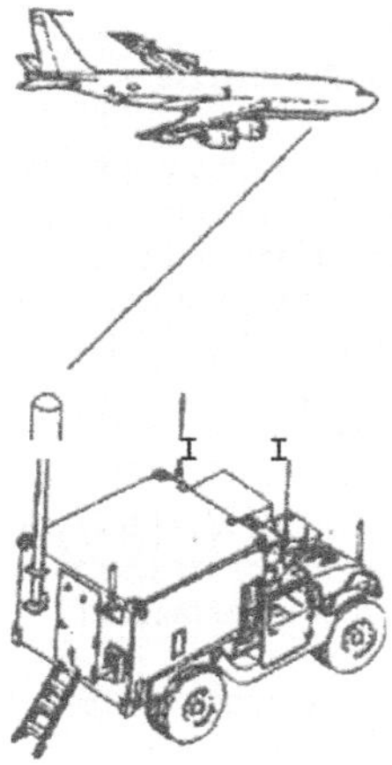

Joint STARS Common Ground Station (CGS)

FUNCTION: MTI, FTI, SAR imagery; supports situation development, battle mgt, targeting, force protection, limited BDA/TDA, operations planning; includes SATCOM & CTT communications

PRIME MOVER: LT - 2xHMMWV

UNIT AND QTY: 5 to 7- DIV; 5 to 7- Corps; 1- ACR; 1- SEP BDE; 1 to 5 EAC

INTEL APPLICATION: IMINT/SIGINT/INTEL/ ARTY/AVN SUPPORT

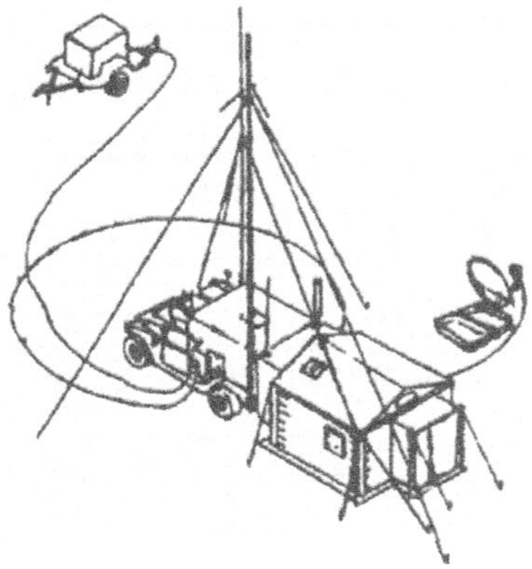

Integrated Meteorological System (IMETS)

FUNCTION: Receives, processes & disseminates weather data.

PRIME MOVER: Heavy HMMWV mounted

UNIT AND QTY: DIV, Corps, EAC, SEP BDE, ACR, AVN BDE, other task organized contingencies

INTEL APPLICATION: Intelligence Support

Advanced Electronic Processing and Dissemination System (AEPDS)

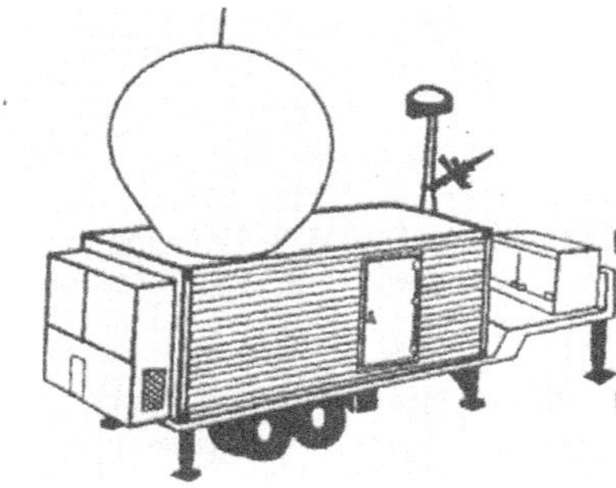

FUNCTION: Mobile, air & ground transportable comm system; receives, processes & exploits SIGINT data from multiple national, theater & organic ELINT/ COMINT systems. Combines functions of EPDS & ETUT into a single downsized system

PRIME MOVER: 30 ft van pulled by a 3-axle 10-ton tractor air deployable on C-130 or larger aircraft

UNIT AND QTV: 1- Corps (rear) 1- Select MI Bde (EAC)

INTEL APPLICATION: SIGINT

Improved Remotely Monitored Battlefield Sensor System (IREMBASS)

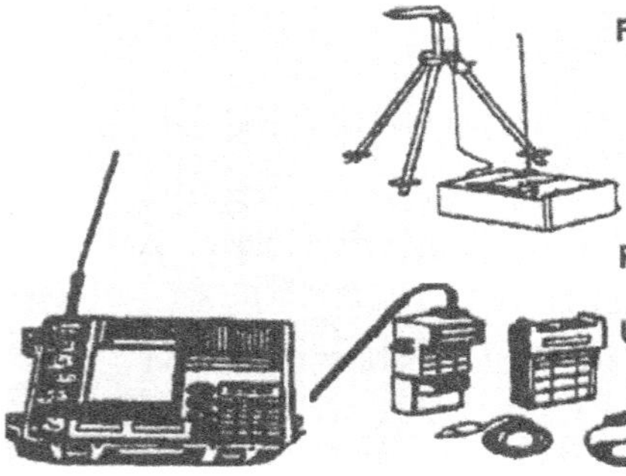

FUNCTION: Unattended ground sensor (seismic/ acoustic, magnetic, infrared, meteorological) that detects, classifies & determines direction of movement

PRIME MOVER: Manpacked and vehicle

UNIT AND QTY: LT DIV - 5 (GS Plt, I&S Co, MI Bn); ASSLT DIV - 5 (GS Plt, I&S Co, MI Bn); ABN DIV - 5 (GS Plt, MI Co (Fwd Spt), MI Bn)

INTEL APPLICATION: MASINT

Mobile Integrated Tactical Terminal (MITT)

FUNCTION: Replaces THMT & FAST-I in force structure; compatible w/TROJAN, MSE, TIBS, DIN/DSSCS, STU III, TENCAP; TRAP/TADIXS-B, UHF SATCOM, S-BAND w/ROTERM/ CHARIOT

PRIME MOVER: Mobile, ground, air transportable (heavy duty HMMWV, cargo HMMWV, generator trailer, C-130, CH-47)

UNIT AND QTY: DIV and Corps ACE - 1

INTEL APPLICATION: SIGINT/IMINT

Tactical Unmanned Aerial Vehicle (TUAV) - OUTRIDER

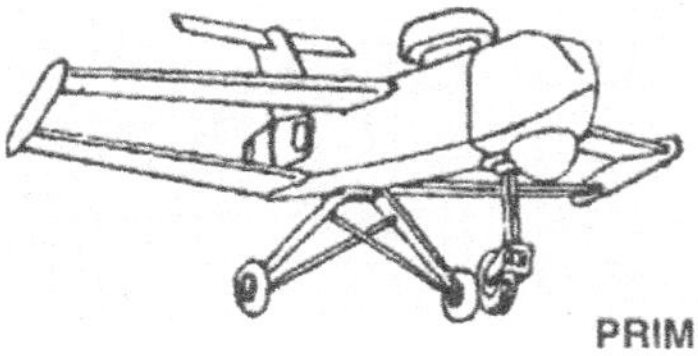

FUNCTION: Operates fwd of FLOT, in all conditions; provides NRT intel, recon, battlefield survl, color E-O camera, FLIR, plug-in optronic payload; future payloads: MTI, SAR, comm/ noncomm DF, intercept, jamming; MET

PRIME MOVER: C-130 or larger, roll-on/roll-off; ground equipment mounted on HMMWVs

UNIT AND QTY: 3 - HVY/AASLT DIV; 5 - LT DIV; 3 - ABN DIV; 3 - ACR; 5 - LT ACR; 1 - SEP BDE

INTEL APPLICATION: IMINT/SIGINT/PSYOP

Unmanned Aerial Vehicle-Short Range (UAV-SR) HUNTER

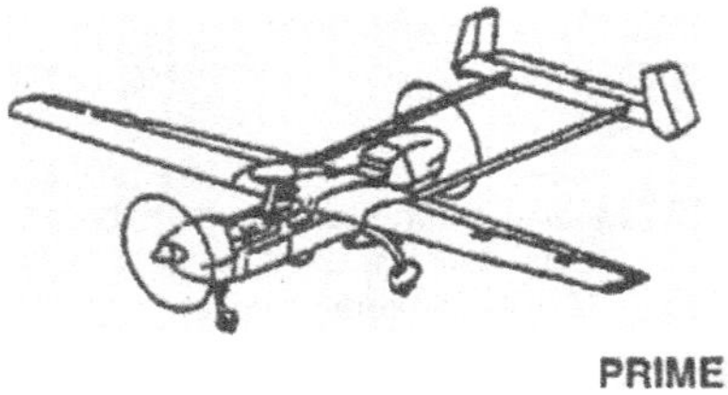

FUNCTION: Operates fwd of FLOT, in all conditions; provides NRT intel, recon, battlefield survl, color CCD, TA, BDA, recon, battlefield survl, E-O/FLIR, UAV data relay; future payloads: MTI, SAR, MET, ES, EA, mine detection, PSYOP, SIGINT, NBC recon

PRIME MOVER: NA

UNIT AND QTY: **Not fielded. For training use only**

INTEL APPLICATION: IMINT/SIGINT/PSYOP

Unmanned Aerial Vehicle (UAV)-Endurance PREDATOR

FUNCTION: Operates as stand-off system in all conditions, behind friendly lines; can penetrate into enemy territory; payloads: E-O/FLIR, MTI/SAR, SIGINT, MET, PSYOP, NBC recon, comm/data relay

PRIME MOVER: 2 C-141s or multiple C-130s

UNIT AND QTY: USAF Asset - Deployed in support of Joint Operations

INTEL APPLICATION: IMINT/SIGINT/PSYOP

Lightweight Man-Transportable Radio Direction Finding System (LMRDFS) AN/PRD-12

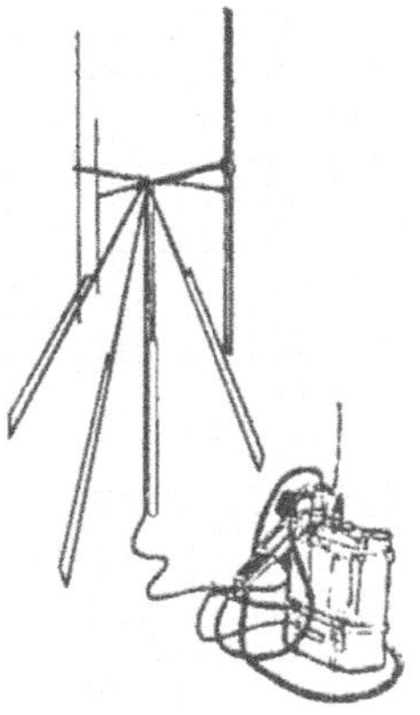

FUNCTION: Lightweight manportable comm intercept/DF system; on emitter location missions, works w/TEAMMATE; ideal in LT, ABN, AASLT, SOF opns; rapid relocation; optimizes collection in fwd AO; CNR for external comm

PRIME MOVER: Manpacked; 60 lbs, carried by 2 soldiers

UNIT AND QTY: 1 - LT ABN; 3 - HVY/AASLT DIV; 3 - ACR; 1 - SEP BDE; 7 - SOF BN; 10 - SOF SA

INTEL APPLICATION: COMINT

Modernized Imagery Exploitation System (MIES)

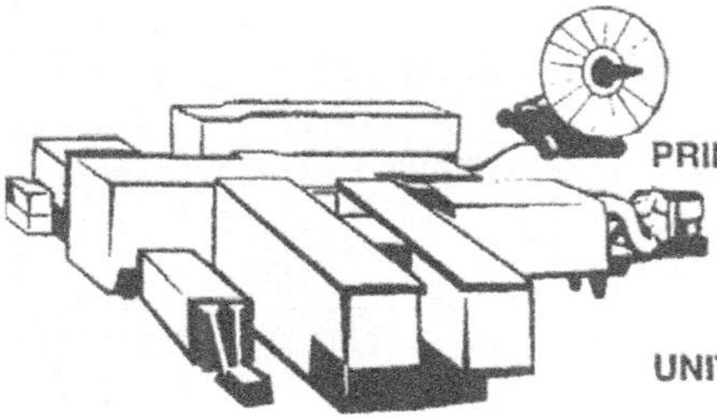

FUNCTION: NRT IMINT; provides exploitation/exploitation mgt, receipt, reporting, SID, archive of imagery & reports; replaces IPDS

PRIME MOVER: 40 ft hardcopy/softcopy exploitation van; 35 ft SATCOM van; 35 ft receive location van; air deployable, C-141, C-5

UNIT AND QTY: Selected MI Bde

INTEL APPLICATION: IMINT

TROJAN SPIRIT (AN/TSQ-190V)

FUNCTION: Satellite terminal; intel processing/dissem, secure voice, data, GPS, FAX, LAN wkstns, video, SID; receive, display, transmit digital imagery, weather/terrain prods among CONUS/OCONUS & deployed

PRIME MOVER: Mounts on 2 heavy variant HMMWVs; tactical SATCOM trailer

UNIT AND QTY: Connectivity: Ft Belvoir TROJAN switching center for worldwide deployment

INTEL APPLICATION: IMINT/SIGINT/CI/ HUMINT/Intelligence Support

---

Synthetic UHF Computer-Controlled Equipment Subsystem (SUCCESS) UHF Radio

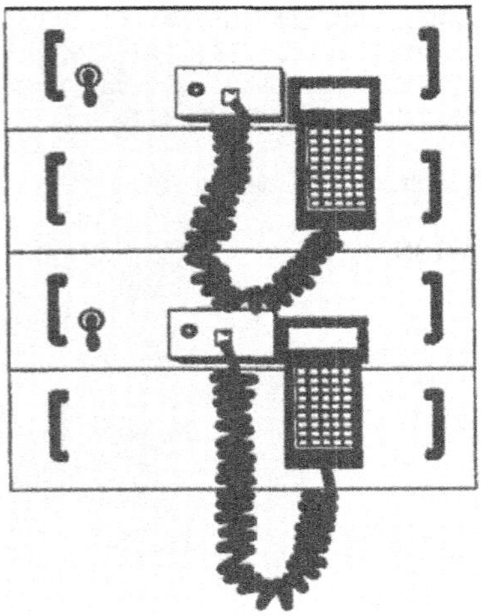

FUNCTION: Auto microprocessor UHF band radio; used w/TENCAP; simultaneous comm when stacked; 2 XMIT/6 RCV channels; SATCOM, point-to-point, TADIXS-B, direct data downlinks; comm w/abn, terrestrial & satellite systems; contains TRE processor

PRIME MOVER: Designed for ground/mobile sheltered environments

UNIT AND QTY: Demand assigned multiple access

INTEL APPLICATION: Intelligence Support

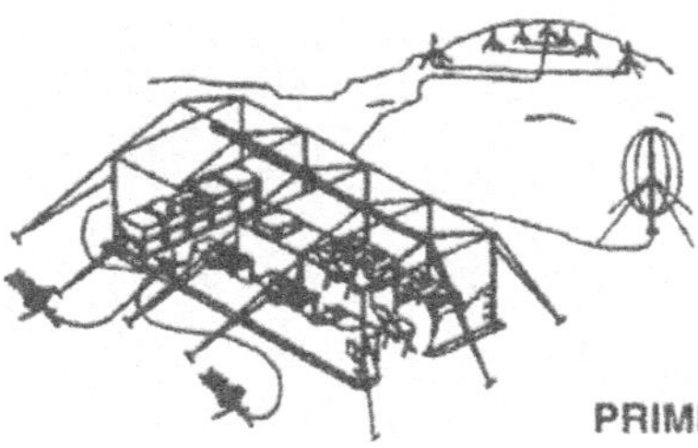

Enhanced TRACKWOLF (AN/TSQ-199)

FUNCTION: Targets HF jammers/emitters; 3 netted or stand-alone stations; produces COMINT, TACREPs; spts EAC w/organic long-distance comm resources; open architecture; scaleable tier I-III; external comm via SINCGARS LPI VHF radio, DIN, DSSCS, TROJAN SPIRIT II, SSP-SIGINT, CROSSHAIR, UHF MIL SATCOM (DF Flashnet)

PRIME MOVER: NA

UNIT AND QTY: Special missions at EAC

INTEL APPLICATION: COMINT

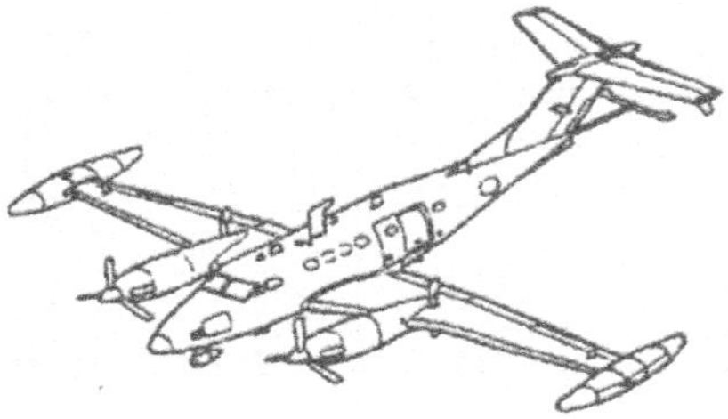

Guardrail Common Sensor (GRCS)

FUNCTION: Collects, identifies, classifies, locates radio signals; provides NRT reporting; gathers LOB & TDOA data; ARF relays platforms btwn IPF & cmds; incorp CHAALS; interfaces w/fixed locations & tactical users; future; expanded coll, LPI capability, embedded trng, 3-channel CTT, retrofit, and auto reporting

PRIME MOVER: Self-deployable; 4 - IPF vans; 3 - IDL trackers

UNIT AND QTY: 1 - Corps AEB

INTEL APPLICATION: ELINT/COMINT

All-Source Analysis System (ASAS)

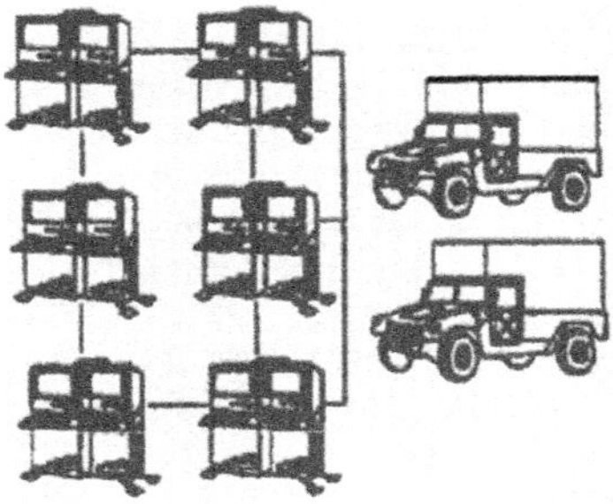

FUNCTION: Provides auto intel, info & interface data handling for cdrs & ATCCS/FLCS users; includes IEW sensors, preprocessors, ASAS & FLCS, ABCS HW/SW & info mgt; RWS operates outside ACE

PRIME MOVER: CUCV/HMMWV 5-ton

UNIT AND QTY: 1 ACE set - EAC, Corps, DIV, Sep Bde, ACR; 1 ASAS-RWS at each Bn, Bde & required collateral locations EAC-DIV

INTEL APPLICATION: COMINT/ELINT/IMINT/CI/HUMINT

Digital Topographic Support System Multi-Spectral Image Processor (DTSS/MSIP)

FUNCTION: Automates terrain TM; produces terrain anal prod; improves response time for IPB; collocated w/ASAS; msg traffic thru MSE; generates/updates multi-spectral imagery; contains 3-D visualization, scanning, map-sizing printer

PRIME MOVER: NA

UNIT AND QTY: 1 DIV Terrain TM; 2 - EAC/Corps Terrain TM

INTEL APPLICATION: IMINT/Intelligence Support

Digital Topographic Support System Quick Response Multicolor Printer (DTSS/QRMP)

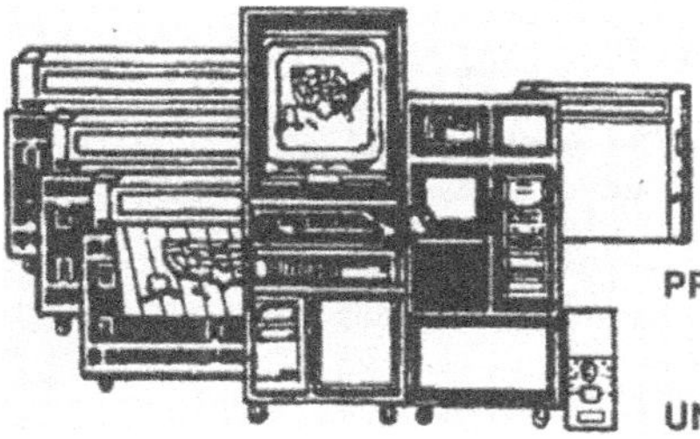

FUNCTION: Automates terrain TM; produces terrain anal prod; improves response time for IPB; collocated w/ASAS; msg traffic thru MSE; provides digital data/terrain anal from battlefield in hard/soft copy

PRIME MOVER: Tactically mobile; 2 LMS-788/gs shelters; 2 HMMWV-H

UNIT AND QTY: 1 DIV Terrain TM; 2 - EAC/Corp Terrain TM

INTEL APPLICATION: IMINT/Intelligence Support

Enhanced Tactical Radar Correlator (ETRAC)

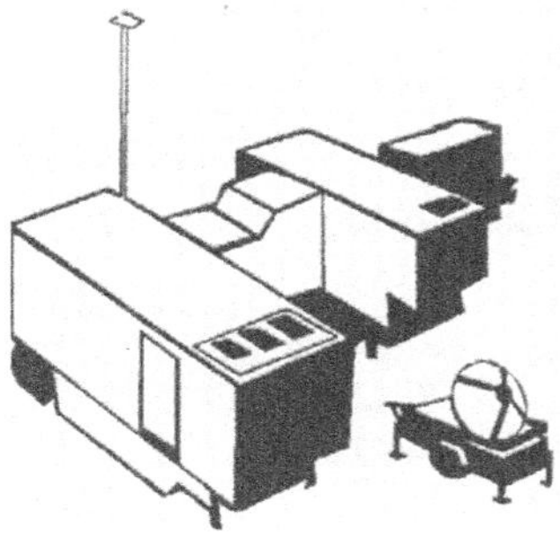

FUNCTION: Advanced mobile SAR processor; receives data from ASAR-2 via U2-R; converts radar to imagery; limited organic exploit for stand-alone opns; includes SUCCESS radio, TENCAP CSP; STU-III, DSVT

PRIME MOVER: Air deployable via C-130, C-141, C-5

UNIT AND QTY: XVIII ABN Corps and V Corps

INTEL APPLICATION: IMINT

CI/HUMINT Automation Tool Set (CHATS)

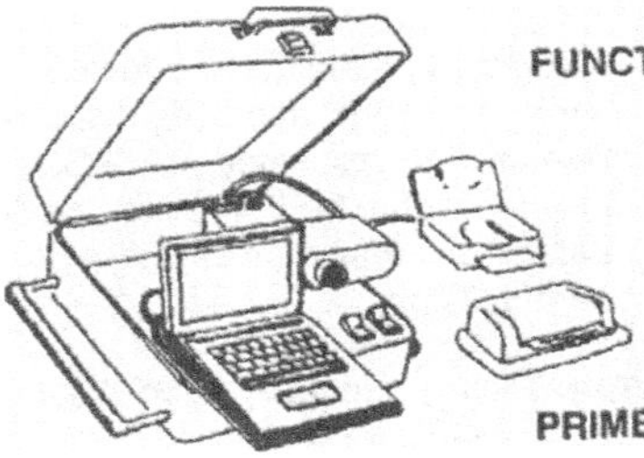

FUNCTION: Provides direct access into CI/HUMINT collateral LAN; uses INMARSAT, TROJAN SPIRIT, SINCGARS, MSE, DSN & phone lines; 2-way data comm; secure FAX; future: GPS, cellular comm, power mgt

PRIME MOVER: 2-manportable cases

UNIT AND QTY: 66th MI Group, Augsburg, GE

INTEL APPLICATION: CI/HUMINT

AN/PPS-5B

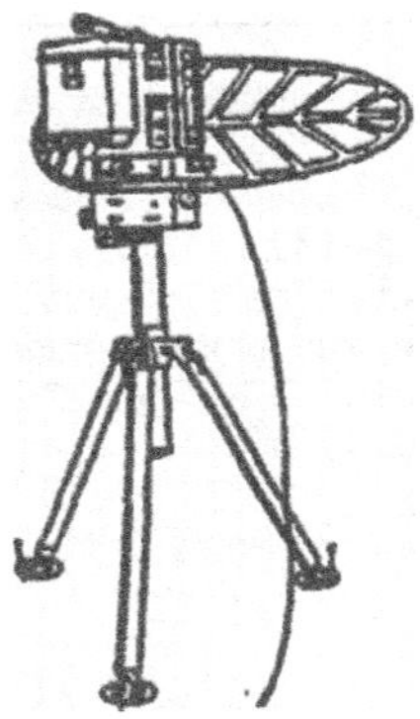

FUNCTION: Moving Target Indicators
Range:
6 km-Personnel
10 km-Vehicle

PRIME MOVER: Manpacked and vehicle

UNIT AND QUANTITY: HVY DIV: 12 systems
4/GSR Squad (3)
Survl Plt, I&S Co., MI Bn

AN/PPS-15A

FUNCTION: Moving Target Indicators
Range:
1.5 km-Personnel
3.0 km-Vehicle

PRIME MOVER: Manpacked and vehicle

UNIT AND QUANTITY: LT DIV: 12 systems
3/GSR Squad (4)
Survl Plt, I&S Co., MI Bn

ASSAULT DIV; 8 systems
3/GSR Squad (3)
Survl Plt, I&S Co., MI Bn

ABN DIV: 9 systems
3/I&S Plt (3)
MI Co., (Fwd Spt), MI Bn

APPENDIX H

MILITARY INTELLIGENCE UNITS AND CAPABILITIES

This appendix shows the structure of the MI Battalion (Heavy Division), the MI Battalion (Light Divisions, Airborne, Air Assault), and the Corps MI Brigade. The MI unit structures constitute approved MI force design.

DIVISION MI BATTALION

The MI Battalion (Heavy Division) (Figure H-1) provides dedicated IEW support to the heavy division. It responds to missions assigned by the commander and staff. (MI battalion IEW assets are shown in Appendix G.)

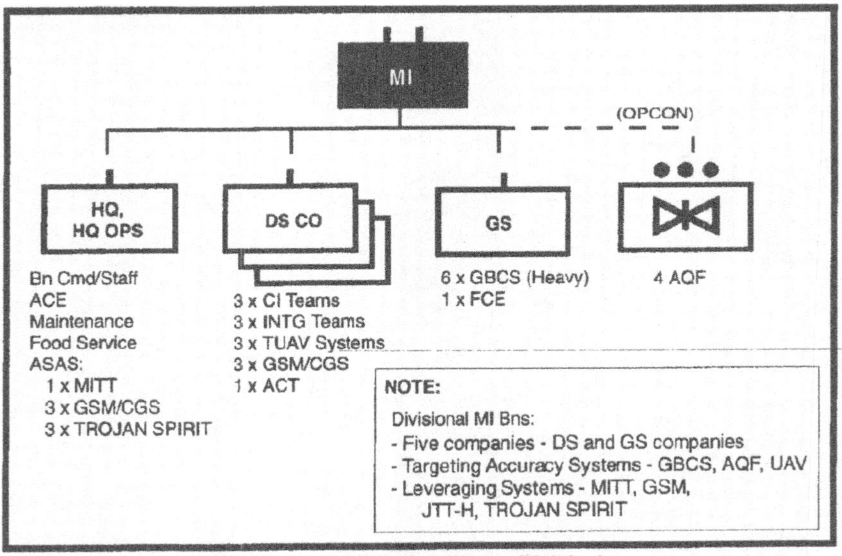

Figure H-1. MI Battalion (Heavy Division).

LIGHT, AIRBORNE, AND AIR ASSAULT DIVISION

The MI Battalion is organized as shown in Figure H-2.

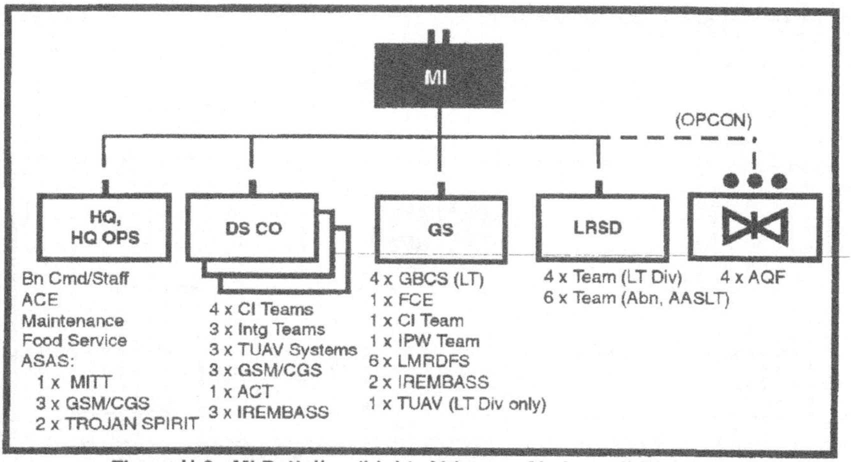

Figure H-2. MI Battalion (Light, Airborne, Air Assault Division).

CORPS MI BRIGADE:

The Corps MI Brigade (Figure H-3) provides dedicated IEW support to the Corps and its subordinate units.

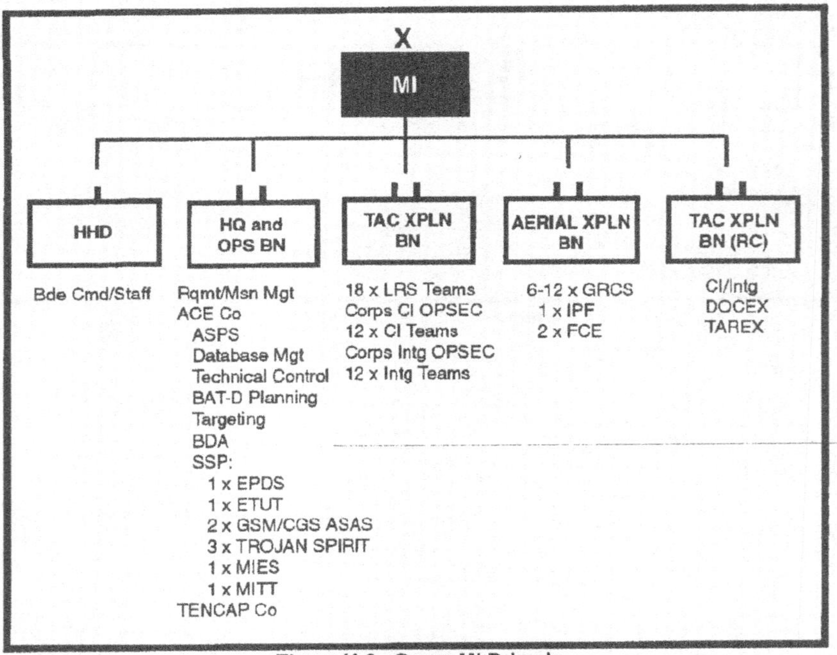

Figure H-3. Corps MI Brigade.

Figures H-4 and H-5 show the organizations of the MI Detachment, Support Company, Special Forces Group and the Special Forces Battalion, respectively.

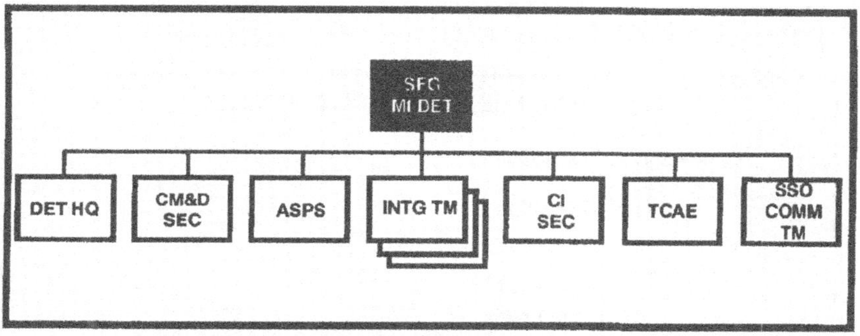

Figure H-4. Organization, MI Detachment, Spt Co, SFG.

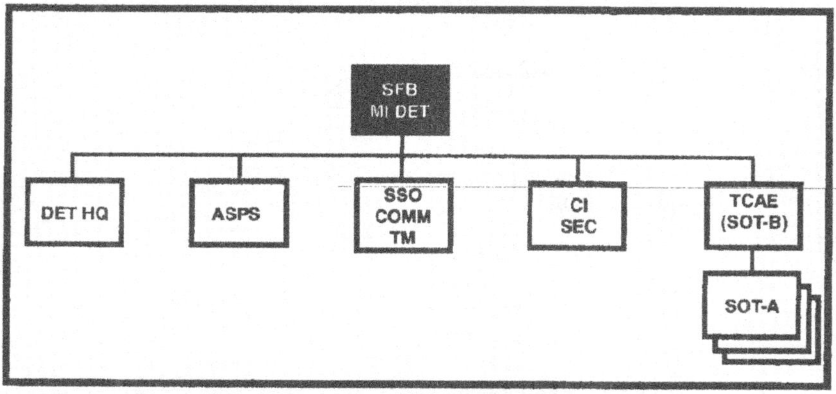

Figure H-5. Organization, MI Detachment, Spt Co, SFB.

APPENDIX I

INTELLIGENCE TRANSITION CHECKLIST

This checklist provides for transition of intelligence responsibility for an area between units. Status of all items listed must be thoroughly coordinated between the transitioning S2s to ensure an effective hand-off of responsibility.

INTELLIGENCE TRANSITION CHECKLIST

_____ CURRENT SITUATION.

- _____ Threat disposition:
  - _____ Maneuver.
  - _____ Artillery.
  - _____ Counter-Air.
  - _____ Intelligence.
  - _____ Mobility-Countermobility-Survivability.
  - _____ Logistics.
  - _____ C^3.
  - _____ Deception.
- _____ Threat strength:
  - _____ Percentage strength of units.
  - _____ BDA.
    - _____ Personnel.
    - _____ Equipment.
    - _____ Weapons.
    - _____ Logistics.
- _____ Threat vulnerabilities.
- _____ Threat capabilities.
- _____ Probable COA.

_____ FRIENDLY INTELLIGENCE SITUATION.

- _____ Collection plan.

_____ IMINT.
  _____ Required coverage.
  _____ Focus (CCIRs).

_____ SIGINT.
  _____ Required coverage.
  _____ Focus (EEFI/PIR).

_____ HUMINT.
  _____ NAIs.
  _____ Surveillance.
  _____ CI/IPW.
  _____ DOCEX.
  _____ Focus (RFI/PIR).

_____ Collection asset status.
  _____ Assets.
  _____ Asset schedule - previous 24 hours next 24 hours.

_____ POCs.
_____ Downlinks and COMINT.
_____ Reporting times (as of time to report and due times).
_____ COS/DEA/DOS.

_____ Lists of stay-behind elements.
  _____ Intelligence assets.
  _____ Communications architecture and frequencies.
  _____ CIA.
  _____ DIA national MIST.
  _____ JCMEC and FMIB.

_____ SIGINT.
  _____ Technical data (adds, changes, deletes).
  _____ MCSF.

_____ EPW.
  _____ Numbers, locations, Category A, SOFA.
  _____ Handover to JTF or ARFOR.
  _____ HN handover.

_____ HN Intelligence/security.
  _____ Support, assets, communications.
  _____ POCs and LNOs.

_____ KEY TERRAIN.

_____ MAP UPDATES AND CORRECTIONS.

_____ DETAINABLE, OF INTEREST, AND PROTECTABLE LISTS.

_____ INTELLIGENCE COLLECTION PLAN.

_____ COORDINATION FOR STAY-BEHIND PERSONNEL.

_____ COVER, SECURITY, OPSEC STATUS, AND PLAN.

_____ LIST OF STAY-BEHIND PRODUCTS:

_____ Intelligence map with current situation and sources.

_____ All-source intelligence support and sources.

_____ Imagery products and sources.

_____ Current weather forecast and on-hand weather, light data, and sources.

_____ Historical documents, locations, and sources.

_____ ENCLOSURES:

_____ List of threat, strengths, capabilities, COAs, weaknesses and vulnerabilities, and effects of friendly action.
_____ Threat situation overlays.
_____ Troop list of friendly collectors.
_____ Friendly collector location overlays.
_____ Intelligence communications net diagrams and POC list.
_____ Key terrain overlays.
_____ Map updates and corrections (may be overlay).
_____ Detainable, of interest, and protectable lists.
_____ Intelligence collection plan.

APPENDIX J

INTELLIGENCE REFERENCE LIST

NOTE: This list is not all inclusive. Include the series of field manuals for your particular unit; for example, FM 44-series for Air Defense. Make sure the publications clerk is aware of your requirements. Refer also to the References section in this manual.

AR 190-11	Physical Security of Arms, Ammunition and Explosive
AR 190-13	The Army Physical Security Program
AR 190-14	Carrying of Firearms and Use of Force for Law Enforcement and Security Duties
AR 190-51	Security of Unclassified Army Property
AR 525-13	The Army Combating Terrorism Program
AR 380-5	Department of the Army Information Security Program
AR 380-10	Technology Transfer, Disclosure of Information and Contacts with Foreign Representatives
AR 380-19	Information Systems Security
AR 380-67	The Department of the Army Personnel Security Program
AR 381-1	Security Controls on Dissemination of Intelligence Information
AR 381-10	US Army Intelligence Activities
AR 381-12	Subversion and Espionage Directed Against the US Army (SAEDA)
AR 381-20	The Army Counterintelligence Program
(U) AR 381-172	Counterintelligence Force Protection Source Operations (CFSO) and Low Level Source Operations (LLSO), Secret.
AR 530-1	Operations Security (OPSEC)
FM 1-111	Aviation Brigades
FM 5-30	Engineer Intelligence
FM 5-36	Route Reconnaissance and Classification
FM 6-20	Fire Support in the AirLand Battle
FM 6-20-10	Tactics, Techniques, and Procedures for the Targeting Process
FM 6-121	Tactics, Techniques, and Procedures for Field Artillery Target Acquisition
FM 7-98	Operations in a Low-Intensity Conflict

FM 11-50	Combat Communications Within the Division (Heavy and Light)
FM 17-95	Cavalry Operations
FM 19-1	Military Police Support for the AirLand Battle
FM 19-30	Physical Security
FM 25-100	Training the Force
FM 25-101	Battle Focused Training
FM 33-1	Psychological Operations
FM 34-1	Intelligence and Electronic Warfare Operations
FM 34-2	Collection Management and Synchronization Planning
FM 34-2-1	Tactics, Techniques, and Procedures for Reconnaissance and Surveillance and Intelligence Support to Counterreconnaissance
FM 34-3	Intelligence Analysis
(U) FM 34-5	Human Intelligence and Related Counterintelligence Operations, Secret.
FM 34-7	Intelligence and Electronic Warfare Support to Low-Intensity Conflict Operations
FM 34-8	Combat Commander's Handbook on Intelligence
FM 34-10	Division Intelligence and Electronics Warfare Operations
FM 34-13	Military Intelligence (MI) Battalion, Combat Electronic Warfare Intelligence (CEWI) (Div), Leaders Handbook for Platoons and Teams
FM 34-25	Corps Intelligence and Electronic Warfare Operations
FM 34-25-1	Joint Surveillance Target Attack Radar System (JOINT STARS)
FM 34-25-3	All-Source Analysis System and the Analysis and Control Element
FM 34-36	Special Operation Forces Intelligence & EW Operations
FM 34-37	Echelons Above Corps (EAC) Intelligence & Electronic Warfare (IEW) Operations
(U) FM 34-40	Electronic Warfare Operations, Secret.
(U) FM 34-40-3	Tactical Signals Intelligence (SIGINT) Analysis Operations, Confidential.
FM 34-40-7	Communications Jamming Handbook
FM 34-40-9	Direction Finding Operations
(U) FM 34-40-12	Morse Code Intercept Operations, Confidential.
(U) FM 34-40-13	Electronic Intelligence (ELINT) Operations, Secret.
FM 34-52	Intelligence Interrogation
FM 34-54	Battlefield Technical Intelligence
FM 34-60	Counterintelligence
FM 34-80	Brigade and Battalion Intelligence and Electronic

	Warfare Operations
FM 34-81	Weather Support for Army Tactical Operations
FM 34-81-1	Battlefield Weather Effects
FM 34-82	Military Intelligence Unit Exercise Development Guide
FM 34-130	Intelligence Preparation of the Battlefield
FM 41-10	Civil Affairs Operations
FM 71-3	Armored and Mechanized Infantry Brigade
FM 71-100	Division Operations
FM 90-2	Battlefield Deception
(U) FM 90-2A	Electronic Deception, Secret.
FM 90-14	Rear Battle
FM 90-29	NEO
FM 100-5	Operations
FM 100-6	Information Operations
FM 100-19	Domestic Support Operations
FM 100-20	Military Operations in Low-Intensity Conflict
FM 100-23	Peace Operations
FM 100-25	Doctrine for Army Special Operations Forces
FM 101-5	Staff Organization and Operations
FM 101-5-1	Operational Terms and Graphics
TC 34-55	Imagery Intelligence
Joint Pub 1	Joint Warfare for the US Armed Forces
Joint Pub 1-01 series	Joint Publication System, Joint Doctrine and Joint Tactics, Techniques, and Procedures Development Program
Joint Pub 2-0 series	Joint Doctrine for Intelligence Support to Operations
Joint Pub 3-0 series	Doctrine for Joint Publications
Joint Pub 5-0 series	Doctrine for Planning Joint Operations
(U) Joint Pub 6-02	Joint Doctrine for Employment of Operational/ Tactical Command, Control, Communications, and Computer Systems, Secret.

NOTE: DOCTRINE WRITTEN PRIOR TO 1992 IS BEING REVISED. UPDATE YOUR REFERENCE LIST AS NEW DOCTRINE IS PUBLISHED.

APPENDIX K

WEATHER ELEMENTS AND SUPPORT

A working knowledge of weather elements is essential to provide your commander with a complete picture of potential adverse weather effects on the battlefield.

INTERACTION OF BATTLEFIELD ENVIRONMENT ELEMENTS

Weather conditions, terrain, BIC, illumination, and background signatures are some primary conditions found in a battlefield environment (Figure K-1).

A weather element is an atmospheric variable that is measured in a weather observation. A weather parameter is derived from one or more weather elements. Example: Density altitude is a weather parameter derived from the element's barometric pressure and temperature.

Weather conditions such as wind, precipitation, and clouds can impact or can be influenced by the other conditions of the battlefield. All of these conditions are interdependent, to a certain degree, and must be considered as a whole (Figure K-2).

WEATHER EFFECTS AND TERRAIN

Terrain features affect weather elements such as visibility, temperature, humidity, precipitation, winds, clouds. The most common example of terrain affecting weather is that on the windward side of high terrain, such as mountains, the rainfall rate will be greater than on the leeward (opposite) side.

Weather conditions have a definite effect on the terrain and can enhance or limit military operations such as trafficability, water crossing (fording), and the first-round accuracy of supporting FA fires. The responsibility for determining mobility and counter-mobility is given to the terrain team at division.

ATMOSPHERIC PRESSURE: Atmospheric pressure is the pressure exerted by the atmosphere at a given point and measured by a barometer in inches of mercury (Hg) or in millibars (Mb).

Pressure Altitude: This is indicated in an altimeter when 29.92 is set in the barometric scale window. High pressure altitude is critical to the lift capability of fixed wing aircraft.

Density Altitude: This is a place in the atmosphere corresponding to a particular value of air density. High pressure is critical to helicopter operations.

CLOUDS: The amount of sky covered by clouds is usually described in eighths: Overcast (8/8ths), broken (5 to 7/8ths), or scattered (1 to 4/8ths). Cloud conditions are described by the amount of cloud cover and the height of the base of the cloud AGL. A cloud ceiling is the height of the lowest broken or overcast layer, and is expressed in feet. A higher layer of several scattered layers of clouds is designated as a cloud ceiling when the sum of the coverage of the lower layers exceeds 4/8ths.

DEW POINT TEMPERATURE: Dew point is the temperature to which the air must be cooled for the air to become saturated and allow dew, and probably fog, to form.

HUMIDITY: This is the state of the atmosphere with respect to water vapor content. It is usually expressed as:

Relative Humidity: The ratio between the air's water content and the water content of saturated air.

Absolute Humidity: The measure of the total water content in the air. It is high in the tropical ocean areas and low in the arctic.

PRECIPITATION: Precipitation is any moisture falling from a cloud in frozen or liquid form. Rain, snow, hail, drizzle, sleet, and freezing rain are common types. The intensity of precipitation is described as light, moderate, and heavy.

Light Rain: Drops are easily seen, very little spray, and puddles form slowly, and accumulation is a trace to 0.10 in/h.

Figure K-1. Common weather elements.

Moderate Rain: Drops are not easily seen, spray noticeable, puddles form rapidly, and accumulation rates are 0.11 to 0.30 in/h.

Heavy Rain: Drops are not seen, rain comes in sheets with heavy spray, puddles form quickly, and the rate is more than 0.30 in/h.

Light Snow: Visibility is equal to or greater than 5/8 miles, or 1,000 meters in falling snow; and a trace to 1 in/hr accumulates.

Moderate Snow: Visibility is 5/16 to ½ statute miles, or 500 through 900 meters in falling snow with 1 to 3 in/hr accumulation.

Heavy Snow: In heavy snow, visibility is cut to less than 1/4 statute miles, or 400 meters, with more than accumulation of 3 in/h.

TEMPERATURE: Temperature is the value of heat or cold recorded by a thermometer normally at 6 feet AGL at the observation site. Temperatures are normally given in both Fahrenheit and Celsius values. It is sometimes referred to as the ambient air temperature.

VISIBILITY: A measurement of the horizontal distance at the surface or aloft that the unaided human eye can discern a large object or terrain feature. Visibility is reported in meters or fractions of a mile, and is reported as a prevailing value of the visibility in all directions. Thus, a visibility report of 1,600 meters may not reveal that fog is diminishing visibility to 400 meters in the northwest if the observer has good visibility in other directions. However, such an event would typically be carried in the weather observation's remarks section.

WIND SPEED AND DIRECTION: These two measure the rate of movement of the air past a given point and the direction from which the wind is blowing. A gust is a rapid fluctuation in wind speed with a variation of 10 knots or more between peak and lull. Gust spread is the instantaneous difference between a peak and a lull and is important for helicopter operations.

Figure K-1. Common weather elements (continued).

WEATHER ELEMENTS	THRESHOLD VALUES	IMPACTED SYSTEMS
VISIBILITY	1.0 km 3.0 km	•DRAGON, Machine guns. •Main gun, TOW, CAS, thermal viewers.
CLOUDS	500 ft 1,000 ft 1,500 ft 5,000 ft	•NOE operations, airborne, helicopter TA. •CAS. •COPPERHEAD engagements. •Visual reconnaissance, ADA TA.
SURFACE WIND	6 kn 13 kn 18 kn 30 kn	•Chemical. •Airborne-round parachute. •Military free fall-ram air parachutes, artillery smoke loses its effectiveness, artillery fire accuracy. •Helicopter maneuver.
TEMPERATURE	>32°C < 0°C	•Helicopter lift capabilities. •NVG (PVS-5).
WINDCHILL	<10°C	Personnel survival.
PRECIPITATION	Moderate Freezing	•Ground maneuver, chemical, laser systems, GSR. •Ground maneuver.

Figure K-2. IPB weather overlay criteria.

BATTLEFIELD-INDUCED CONTAMINANTS

During combat operations, visibility can become severely reduced by BIC. These contaminants are either induced directly by combatants or occur as by-products from battlefield operations.

Two significant sources of battlefield contaminants are dust produced by HE artillery or mortar rounds and deliberately employed smoke. HE rounds usedin a pre-attack barrage not only may kill enemy forces but also may restrict the visibility of your own troops by dust caused by the HE if you do not consider the direction of the wind. Smoke produced by smoke generators, vehicle exhaust emission systems, smoke pots, indirect fire, and smoke rounds also produce battlefield contaminants.

Wind speed and direction are critical to maintaining an effective smoke screen. Rain can remove BIC quickly. Weather inversions over valley areas can sustain airborne contaminants for long periods.

Other sources of BIC that will lower visibility in the AO are clouds of dust from vehicle traffic or smoke from fires. These types of contaminants not only blind you but also may help your adversary in detecting troop movements and pinpointing your location.

A unique BIC affecting visibility occurs in conditions when temperatures are in the range of -30°C or -22°F or colder. When a source of moisture or water vapor is released into the cold air by internal combustion engines, artillery fires, or launched self-propelled munitions, visibility can be reduced to zero when the moisture freezes instantly and changes into ice fog.

Ice fog may restrict visibility across a whole valley and can linger for hours. Ice fog crystals permit ground objects to be seen from above while severely restricting visibility on the ground—an advantage for aerial reconnaissance.

On airfields an ice fog created by fixed-wing aircraft may cover an entire runway. Visibility can be reduced so that other aircraft cannot take off or land if the wind is calm. The ice fog also draws attention to the airfield location.

Launching missiles such as the TOW in very cold air can create an ice fog. As the TOW moves to the target, the exhaust blast exits into the air where it condenses and creates the ice fog. If the wind is calm, this fog follows the trajectory of the missile and reduces launch point visibility to such an extent that the operator loses sight of the target. Also, the launch point can be identified by threat forces from the condensation trail of the missile.

ILLUMINATION

Natural light is critical in planning operations where NVD are used or in operations timed to use only available light. Natural light values vary as a function of the position of the sun, moon, stars, and clouds. Light data are available from your SWO for any time period and place. These data are particularly important for determining first and last light, moonrise, and moonset, and are most effective for planning use of NVD.

Variables such as altitude, cloud cover, terrain-produced shadows, visibility, and direction of vehicle or aircraft movement in relation to the sun or the moon can also affect light level availability.

Artificial light is intended to increase visibility but, under certain weather conditions, this does not always occur. Example: Low cloud ceilings will limit the area covered and effective time of flares. Rain, snow, or fog can reduce flare effectiveness. However, under the right conditions, cloud cover can enhance the effects of artificial light due to cloud base reflection. Snow- or sand-covered terrain also reflect both natural and artificial light.

BACKGROUND SIGNATURES

Temperature, wind, and precipitation have a major influence on your ability to pick out a target from the background in the infrared spectrum. They also affect seismic (sound and acoustic) signatures. Detection of objects in the infrared spectrum depends on a temperature contrast between the object and its surrounding environment. This difference is known as the background signature.

Snow, rain, and wind influence the background signature because they can change the surface temperature of objects. These elements lower object temperatures and thus reduce the differential between a target and its background.

A heavy layer of snow produces a washout during any part of the day since it causes both the object and the background to exhibit the same temperature.

Precipitation also degrades seismic sensors through the introduction of background noise (rain), while a snow-covered surface will dampen sound and the movement of troops.

SEVERE OR HAZARDOUS WEATHER

In addition to a continuous need for forecast updates, you need non-forecasted or unanticipated severe or hazardous weather warnings. WETMs normally issue severe weather warnings and advisories. Check for the values at which each weather element becomes known as severe. You will be concerned with conditions such as—

- Tornadoes.
- Thunderstorms producing winds in excess of 45 knots and hail greater than 3/4 of an inch.
- Hurricanes and typhoons.
- Precipitation (rain or snow) when X inches fall in Y hours.
- Surface winds in excess of X knots.
- Maximum and minimum temperatures; when a forecast value misses the actual temperature by X degrees.

You want to know that an earlier forecast for light snow was amended to a forecast of a 16-inch accumulation within the next 12 hours. We cannot over emphasize that you need to work with your next higher HQ S2/G2 and the SWO so that your needs are realistically stated and can be supported. All of your weather support needs should be reviewed every 6 months in garrison and as required in the field.

LIGHT DATA

Another weather-related element that your commander needs is light data. NVD and NVG have made many night operations feasible. Your SWO provides official times for sunrise and sunset, BMCT and EECT, BMNT and EENT, length of absolute darkness, moonrise, moonset, lunar phase in percent, and time periods for using NVD and NVG. Light data to support NVD is needed because there are times when there is not enough moon or starlight to use them.

Civil twilight is sufficient for conduct of combat operations while nautical twilight permits most ground movements without difficulty. Nautical twilight allows a general visibility of up to 400 meters (1,320 ft) and lets you distinguish silhouettes from the background.

The actual duration of light varies with latitude and time of year. For example, in the vicinity of 35 to 40 degrees north latitude, civil twilight generally occurs 30 to 45 minutes before sunrise and after sunset. In the tropics, twilight is shorter.

Once light requirements are determined, relay them to your next higher S2 and SWO. This information is important for your commander because he needs to know not only when he can begin friendly military operations (day or night) but also when threat operations could begin.

OTHER CONSIDERATIONS

Although HF radio wave propagation forecasts are not normally available to the SWO, he can make arrangements before he deploys to receive these forecasts. When available, they should be given to every signal and intelligence organization. The SIGO should know that when HF is not effective, it may be because of solar activity rather than enemy jamming.

The USAF Global Weather Central can routinely provide solar forecast products to the SWO even in the field. Other data, such as tidal information and sea state conditions, can be obtained from the SWO but are normally provided only to specific Army units upon request.

RECEIPT OF FORECASTS

Every tactical echelon should receive the weather forecast prepared and briefed to the commander at the next higher echelon. With no SWO at a maneuver brigade, you receive both the forecast briefed at division covering the division AO and the forecast made by the division SWO specifically for your brigade. Each forecast message received should be worked by you to discover the direct weather impacts on your unit.

A commander wants the weather forecast. He also needs the effects and impacts of the forecasted weather interpreted for his specific operations, systems, and personnel. Schedule your forecasts so they arrive in time for you to prepare your commander's briefing.

HOW G2 OR S2 RECEIVE WEATHER SUPPORT

It is your responsibility to provide weather and weather effects information to your commander and supported or subordinate units. Methods may vary among units and echelons. At higher echelons, the SWO has primary responsibility for providing weather support. There, the G2 simply contacts the SWO, states requirements, and receives the needed weather support.

If you have no SWO, you must pass the request up the chain of command to a level where one is assigned. Once your requirements have been validated, weather support products flow back over the same path. In addition to weather forecast products, the SWO can provide specialized weather effects products. You must provide the threshold values used in developing these products.

You alert your commander and staff about the adverse impact of the forecasted weather. You prepare two simple charts to use during your commander's stand-up briefing. You can post these charts in the CP for reference.

Update the charts when a new forecast is issued by the WETM. Methods used in briefing your commander during the morning and evening briefings may vary greatly among commands. The formats below are a guide and should be modified to suit your situation.

The first chart (Figure K-3) illustrates how the weather elements and parameters contained in the forecast might be displayed on the board. Blown up to poster size and covered with acetate, this chart can be updated easily.

WEATHER FORECAST VALID FOR 121200Z to 131200Z	
LOCATION: PUNGSAN	
24-HOUR FORECAST:	
SKIES:	CLEAR MORNING AND NIGHT, PARTLY CLOUDY IN THE AFTERNOON, BASES 3,000 FEET.
VISIBILITY:	UNLIMITED, OCCASIONALLY 1 TO 2 MILES IN BLOWING SNOW DURING AFTERNOON.
WINDS:	NORTH TO NORTHWEST, 10 TO 15 KNOTS, OCCASIONAL GUSTS TO 25 KNOTS IN AFTERNOON.
TEMPERATURES:	MAX: 10° F MIN: -20°F
72-HOUR OUTLOOK:	CLOUDY SKIES, SNOW FLURRIES DURING AFTERNOON HOURS LOWERING VISIBILITY TO 2 TO 4 MILES. TEMPERATURES: MAX: 20°F MIN: -5°F
LIGHT DATA:	BMNT: 1247Z MOONRISE: 1924Z BMCT: 2221Z MOONSET: 0819Z SUNRISE: 2251Z SUNSET: 0820Z NVG USE: EECT: 0851Z

Figure K-3. Example of a weather forecast chart.

Figure K-4 shows color codes as one way to display potential weather impacts on operations, systems, and personnel in your unit. Another way would be to write the words "moderate" and "severe" in those blocks affected. Do not list all the equipment or systems, but have the list available to answer questions posed by the commander or staff. Stress those critical systems during the verbal portion of your briefing.

If the weather conditions change significantly during the period covered by the SWO's forecast, then an updated impact chart will have to be prepared. Because a brigade or battalion's AI is small, the SWO's forecast is likely to be uniform across the AI.

WEATHER IMPACT CHART

PERIOD 121200Z to 131200Z

IMPACTED ITEMS	CLOUDS	PRECIPI-TATION	WIND	VISIBILITY	TEMPER-ATURE
NBC DEFENSE	GREEN	GREEN	GREEN	GREEN	RED
NIGHT VISION DEVICES	GREEN	GREEN	GREEN	AMBER EXCEPT 02 to 05Z	AMBER
CSS	GREEN	AMBER	AMBER	AMBER	AMBER
TOW/DRAGON	GREEN	GREEN	AMBER	GREEN	RED
M-1/M-2	GREEN	GREEN	GREEN	GREEN	AMBER
SMOKE/ CHEMICAL	GREEN	GREEN	AMBER	GREEN	GREEN
GROUND RADAR	GREEN	GREEN	AMBER	GREEN	GREEN
PERSONNEL	GREEN	GREEN	RED	GREEN	RED
THREAT	GREEN	GREEN	GREEN	AMBER	GREEN

GREEN	No impact	AMBER	Moderate impact (Normal effectiveness reduced 25 to 75%)	RED	Severe impact (Normal effectiveness reduced to 0 to 25%)

Figure K-4. Example of a weather impacts display chart.

CRITICAL WEATHER FACTORS

Figure K-5 shows major critical weather factors that can enhance or degrade combat operations, systems, and personnel.

LOW VISIBILITY:	Low visibility (less than 3 km). • Can be beneficial to threat and friendly forces. • Conceals the center of gravity and maneuver of offensive forces. • Increases possibility of achieving surprise. • Disadvantages. –Hinders C^2. –Reduces effectiveness of R&S and TA, especially during the defense.
SURFACE WINDS:	• Strong winds can reduce effectiveness of downwind forces by blowing dust, smoke, sand, rain, or snow on them. • Upwind force generally has better visibility and can advance faster and easier. • Strong winds limit airborne and aviation (primarily helicopters) operations. • Winds in excess of 35 knots can— –Cause personal injury. –Damage materiel and structures. –Create false radar returns. –Reduce visibility because of blowing sand, dust, and other battlefield debris.
PRECIPITATION:	• Affects trafficability, visibility, personnel effectiveness, and a wide variety of tracked and wheeled military equipment. • Heavy rains can make some unsurfaced, low-lying, and off-road areas impassable.

Figure K-5. Critical weather factors.

	• Both rain and snow can drastically reduce personnel effectiveness by limiting visibility, causing discomfort, increasing fatigue, and creating other physical and psychological problems.
CLOUD COVER:	• The type and amount of cloud cover and the altitude of cloud bases and tops influence aviation operations. • CAS employing fixed-wing aircraft would like a ceiling of at least 2,500 feet (762 m), but can be employed with ceilings as low as 500 feet. • Threat CAS rotary-wing aircraft and aerial resupply missions require a minimum ceiling of 300 feet (100 m). • Affects ground operations by reducing illumination and visibility or by enhancing effects of artificial light.
TEMPERATURE AND HUMIDITY:	• Have a direct impact on personnel and vehicle performance. • Excessively high temperatures cause heat-related injuries to personnel and vehicle engine wear that leads to equipment failure. • Very low temperatures increase cold weather injuries, cause damage to vehicle cooling systems and engines, decrease the effectiveness of vehicle lubrication, and create excessive logistics requirements.

Figure K-5. Critical weather factors (continued).

WEATHER EFFECTS ON THREAT SYSTEMS

MILITARY ASPECTS OF WEATHER:

US Forces must be prepared to fight in a variety of climatic conditions on short notice. Key to accomplishing our missions under any circumstances is understanding how weather affects both friendly and threat forces and their operations, systems, and personnel.

Current weather conditions and weather forecasts for the AO and AI are analyzed to determine the effects on friendly and enemy operations. This is significant when threat forces have the capability to employ NBC weapon systems.

THREAT EQUIPMENT:

Some of the major arms merchants today are the former Soviet Union, Sweden, Brazil, Britain, Germany, France, Italy, and the United States. However, the major arms purchasers continue to be the underdeveloped or Third-World countries in the Middle East, Latin America, and Asia.

The types of threat equipment we may encounter on future battlefields will vary from artillery and mortars produced during World War II to the Austrian-produced GHN-45, a 155-mm towed gun with a range of 39,600 meters (using ERFB-BB technology). Almost all the Third-World countries have bought or made their own versions of Soviet-produced tanks, APCs, artillery, AAA, and assorted SAM and SSM systems. Tables K-1, K-2, and K-3 list the weather effects from cloud ceilings, reduced visibility, and precipitation, respectively.

Table K-1. Weather effects from cloud ceilings.

WEATHER VALUE (FEET)	SEVERE DEGRADATION		MODERATE DEGRADATION	
	SYSTEM/ EVENT	REMARKS	SYSTEM/ EVENT	REMARKS
LT 1,000			SA-9 SAM SA-14 SAM	
LT 2,500			ZU-23 ZSU-23-4 SA-13 (Contrast mode) SA-16 SA-19	

Table K-2. Weather effects from reduced visibility.

WEATHER VALUE (METERS)	SEVERE DEGRADATION		MODERATE DEGRADATION	
	SYSTEM/ EVENT	REMARKS	SYSTEM/ EVENT	REMARKS
LT 200			RPG-18 (RL) AKM/AKMS (7.62-mm) TKN-1T infrared periscope	ATW Rifle E-O device
LT 500	SA-14 GREMLIN SAM SA-13 SAM SA-16 SAM SA-19 (FSU) SAM		OU-3GK White infrared search light RPG-7 (GL) PT-76 (5.45-mm) AK-74	E-O device ATW AG/LT Rifle
LT 600			ERYX	
LT 750	SA-9 GASKIN SAM			
LT 800	AT-2 SWATTER AT-3 SAGGER AT-6 SPIRAL	ATGM ATGM ATGM	ASU-85 BMD (73-mm) BMP (73-mm) Sniper rifle, SVD RPK-74, squad AGS-17, GL RPG-16, GL	AG/LT LAV LAV Rifle MG

Table K-2. Weather effects from reduced visibility (continued).

WEATHER VALUE (FEET)	SEVERE DEGRADATION		MODERATE DEGRADATION	
	SYSTEM/ EVENT	REMARKS	SYSTEM/ EVENT	REMARKS
LT 1,000	All Types	ATGM	SPG-9 (73 mm recoilless rifle) SD-44 (85-mm) 14.5-mm KPU hvy MG 7.62-mm PKT MG DShK NSV/NSVT PK Series MT-LB 7.62-mm coaxial machine gun for all tanks	ATW ATW AA AA MG MG LAV MT
LT 1,500			7.62-mm 15.5-mm heavy BTR-50,-60,-70 (14.5-mm) KPVT ACRV M1974 DShK NSV/NSVT T-54, T-55, T-62 (main gun)	MG MG AA/LAV AA LAV MG MG MT
LT 2,000			T-12, MT-12 (100-mm) KPVT BRDM-2 (14.5-mm) T-80, T-72, T-64 (main gun) MATHOGO MILAN 2 RBS-56 SPIGOT	ATW MG LAV MT

Table K-2. Weather effects from reduced visibility (continued).

WEATHER VALUE (METERS)	SEVERE DEGRADATION		MODERATE DEGRADATION	
	SYSTEM/ EVENT	REMARKS	SYSTEM/ EVENT	REMARKS
LT 2,500			SA-7 GRAIL ZU-23 ZSU-23-4	FSU SAM AA AA
LT 3,000			RED ARROW 73 RED ARROW 8 SUSONGP'O AT-3 SAGGER	
LT 4,000			SA-9 GASKIN HOT 2 SWINGFIRE AT-2 SWATTER AT-5 SPANDREL AT-8 SONGSTER AT-10 STABBER SA-13 (contrast mode)	FSU SAM FSU SAM
LT 4,500			MAPATS	
LT 5,000			NIMROD AT-6 SPIRAL AT-11 SNIPER SWIFT (SF)	ATGM
LT 6,000			120-mm, M-1943 S-60 (57-mm)	MO AA
LT 8,000			160-mm, M-160	MO

Table K-3. Weather effects from precipitation.

WEATHER CONDITION	SEVERE DEGRADATION		MODERATE DEGRADATION	
	SYSTEM/EVENT	REMARKS	SYSTEM/ EVENT	REMARKS
Light rain			LOW BLOW (fire control) STRAIGHT FLUSH (12gHz) (tracking) THIN SKIN (height finder)	SAM radar SAM radar SAM radar
Moderate rain			STRAIGHT FLUSH (acquisition)	SAM radar
Heavy rain			LOW BLOW (fire control) THIN SKIN (height finder)	SAM radar SAM radar
	NOTE: The list of AD radars on this page should be considered as a sample of the various other types of threat radars associated with EW operations and tactical and strategic SAM systems. Not listed are a large variety of threat combat equipment (wheeled and tracked) affected by moderate or heavy rain and snow.			

WEATHER AND ENVIRONMENTAL ELEMENTS AND PARAMETERS IMPACTING ARMY SYSTEMS AND OPERATIONS:

Table K-4 is compiled from a study conducted by the US Army Intelligence School as part of the IEWMAA. Each TRADOC branch was asked to survey the impact of weather and environmental elements on their operations, systems, and personnel and rate them as either**essential (E)** or **desired (D)**. Also included are other Army agencies' requirements.

- **E**—Some positive action had to be taken by the user based on severe current or forecasted weather conditions.
- **D**—Some impact, but the degree of impact is uncertain or not mission-threatening.

For each element or parameter listed, at least one responder identified it as essential for one or more of their operations, systems, or personnel.

Some data elements and parameters are known to have an impact, but exact critical thresholds have not been determined. Others cannot be measured or sensed with present technology. But identifying these data elements or parameters now allows for further research in how to collect the raw data (where applicable), determine the frequency of collection, establish data accuracy, and learn other supporting information.

As users and planners continue to learn more about weather and environmental impacts, the misconception that there is a hypothetical Army "all-weather" system or operation fades.

Table K-5 shows windchill factors.

Table K-4. Weather and environmental data elements impacting Army systems and operations.

WEATHER AND ENVIRONMENTAL DATA ELEMENTS IMPACTING ARMY SYSTEMS/OPERATIONS

A = AIR DEFENSE
B = ADJUTANT GENERAL
C = ARMOR
D = AVIATION
E = CHAPLAIN
F = DEFENSE AMMUNITION CENTER
G = ENGINEER
H = FIELD ARTILLERY
I = FINANCE
J = HEALTH SERVICES
K = INTELLIGENCE
L = INFANTRY
M = JUDGE ADVOCATE
N = MILITARY POLICE
O = CHEMICAL
P = ORDNANCE
Q = ORDNANCE M&M
R = PUBLIC AFFAIRS
S = QUARTERMASTER
T = SIGNAL
U = SPECIAL FORCES
V = SOLDIER SUPPORT
W = TRANSPORTATION
X = JOINT READINESS TRNG CENTER

REQUIREMENT DESIGNATORS: E = ESSENTIAL SUPPORT NEEDS D = DESIRABLE SUPPORT NEEDS

SPECIAL GROUP CATEGORIES: [GS] = GROUND STATE [SS] = SEA STATE [SW] = SEVERE WEATHER CRITERIA

DATA ELEMENT	CATEGORY	A	B	C	D	E	F	G	H	I	J	K	L	M	N	O	P	Q	R	S	T	U	V	W	X
1. ALTIMETER SETTING					E						E	E										E			
2. ATMOSPHERIC CONTAMINANTS					E	D	D	E	D		E	E							D			D			
3. ATMOSPHERIC DENSITY									E			E										E			
4. ATMOS. TRANSMISSION COEFFICIENT									E																
5. BAROMETRIC PRESSURE		D		E	E				E		E	E										E			E
6. CLOUD COVER AMOUNT		D		E	E		E	E	E		E	E	E		D	E			D			D	E	E	E
7. CLOUD BASE HEIGHT		E			E		E	E	E		E	E	E	D		E						E		E	E
8. CLOUD TOP HEIGHT					E						E	E										D			
9. DAMAGING WINDS	[SW]	E	E	E	E	E	D	E	E	E	E	E	E	E	E	E	D	E	E	E	E	E	E	E	E
10. DENSITY ALTITUDE					E						E	E										E			E
11. EXTINCTION COEFFICIENT									E			E										D			
12. EXTREME HEAT/COLD	[SW]	E	E	E	E	E	E	E	E	E	E	E	E	E	E	E	D	E	E	E	E	E	E	E	E
13. FLOODING, RIVER STAGE	[GS]	E	E	E	E	E		E	D	E	E	E	E	E	E	E	D	E	E	E	E	E	E	E	E
14. FREEZE/THAW DEPTH	[GS]	D		E	E			E	D		E	E	E		E	D	E	E		E	E	D			E
15. FREEZING PRECIPITATION	[SW]	E	E	E	E	E		E	E	E	E	E	E	E	E	E	D	E	E	E	E	E	E	E	E
16. HEAVY RAIN/SNOW	[SW]	E	E	E	E	E	D	E	E	E	E	E	E	E	E	E	E	D	E	E	E	E	E	E	E
17. HUMIDITY, ABSOLUTE		D			E			E			E	E				E						E			E
18. HUMIDITY, ABSOLUTE, PROFILE		D			E			E			E	E				E						E			E
19. HUMIDITY, RELATIVE							E	E	D		E	E				E				E		D	E	E	E
20. HUMIDITY, RELATIVE, PROFILE									E													D			
21. HURRICANES/TYPHOONS	[SW]	E	E	E	E	E		E	E	E	E	E	E	E	E	E	D	E	E	E	E	E	E	E	E
22. ICE/SNOW DEPTH/COVER	[GS]	D		E	E			E	D		E	E	E		E	D	D	E		E	E	D			E
23. ICING, FLIGHT	[SW]				E						E	E	E									E			E
24. ILLUMINATION		E		E	E			E	E		E	E	E		E		D			D		E	E	E	E
25. IR TARGET/BACKGROUND CONTRAST				E	E				D		E	D										E			E
26. IR THERMAL CONTRAST X-OVER TIME				E	E				D		E	E										E			E
27. IONOSPHERIC DISTURBANCES					D						D	E									E	E			
28. LIGHT DATA		E		E	E		D	E	D		E	E	E		D	D				D	D	E		D	D
29. LIGHTNING/THUNDERSTORMS	[SW]				D		E	E	E		D	E								E		E			E
30. LITTORAL CURRENT	[SS]				D		D	E			D	D	E									E			
31. PRECIPITATION, ACCUMULATION			D	E	E	D	D	E	D	D	E	E	E		E	D	E		D	E	E	E	E	E	D

Table K-4. Weather and environmental data elements impacting Army systems and operations (continued).

DATA ELEMENT	CATEGORY	A	B	C	D	E	F	G	H	I	J	K	L	M	N	O	P	Q	R	S	T	U	V	W	X
32. PRECIPITATION, RATE		E	D		E	D		E	E	D	E	E	E	D	E	E	D		D	E	E	E	E	E	E
33. PRECIPITATION, HAIL SIZE		E			E			E	D		E	E			E		E			E		D			
34. PRECIPITATION, TYPE		E	D	E	E	D		E	E	D	E	E	E	D	E	E	D		D	E	E	E	E	E	E
35. PRESSURE ALTITUDE					E						E	E	E									E			E
36. REFRACTIVE INDEX		E		E	E				D		E	E			D										
37. RESTRICTION TO VISIBILITY					E			E	D		D	E			E	E				E	E	E			E
38. SEEABILITY (MILLIMETER WAVE, INFRARED, ULTRAVIOLET)		E		E	E			E	E		E	E	E		E	D						E		E	D
39. SEVERE WEATHER CONDITIONS	[SW]	E	E	E	E	E		E	E	E	E	E	E	E	E	E	D	E	E	E	E	E	E	E	E
40. SNOW STATE CONDITION	[GS]	D		E	E			E	E		E	E	E		E	D	D	E		E	E	D			E
41. SNOW DRIFT DEPTH	[GS]	D			D			E	D		D	E	E		E		E			D	E	E			E
42. SOIL/GROUND MOISTURE	[GS]	D		E	E		D	E	D		E	E	E		E	D	D	E		D	E	D			E
43. SOIL/GROUND TEMPERATURE	[GS]	D		E	E		D	E	D		E	E	E		E		E		D	D		E	E	E	E
44. SOLAR RADIATION							E				D	D										E			
45. STABILITY INDEX							E					E				E									E
46. STANDING WATER/POOLING	[GS]	D		E	E	D		E	D		E	E	E		E	D	D	E	D	E	E	D			E
47. STATIC ELECTRICITY POTENTIAL					E						E	E													
48. SURF HEIGHT	[SS]				D			E			D	D	E									E		E	
49. SWELL DIRECTION/HEIGHT	[SS]				D						D	D	E									E		E	
50. TEMPERATURE, AIR, SURFACE		E	E	E	E	E	E	E	E	E	E	E	E		E	E	D	E		D	E	E	E	E	E
51. TEMPERATURE, AIR, PROFILE					E		E		E		E	E										D			
52. TEMPERATURE, AIR, UPPER AIR					E				E		E	E										D			E
53. TEMPERATURE, DEWPOINT				E	E				D		E	E										E			E
54. TEMPERATURE, DEWPOINT, PROFILE					E				E		E	E										D			
55. TEMPERATURE, WINDCHILL FACTOR			D		D	D		E	D	D	E	E	E	D	E		E		D	E		E	E	E	
56. TEMPERATURE, INVERSION LEVEL(S)					E		E		E		E	E				E						D			E
57. TEMPERATURE, SEA SURFACE	[SS]				D			E			D	D	E									E		E	
58. TEMPERATURE, WATER, INLAND	[GS]				D			E			D	E	E									E		E	
59. TEMPERATURE, WBGT					D			E	D		E	E			E	E	E			E		E			E
60. TORNADO	[SW]	E	E	E	E	E		E	E	E	E	E		E	E	E	D	E	E	E	E	E	E	E	E
61. TURBULENCE, FLIGHT					E						E	E										E			
62. TURBULENCE, OPTICAL				E	E			E	E		E	D										E			
63. VISIBILITY, VISIBLE SPECTRUM, ALOFT					E				E														E		E
64. VISIBILITY, VISIBLE SPECTRUM, SFC		E			E			E	E		E	E	E		E	E	D			E		E	E		E
65. WAVE PERIODICITY	[SS]				D						D	D	E									E		E	
66. WAVE DIRECTION/HEIGHT	[SS]				D						D	D	E									E		E	
67. WIND, PROFILE					E				E		E	E										E			
68. WIND, PROFILE, TARGET, AGL - 3KM					E				E							E									
69. WIND, SHEAR					E						E	E										E			E
70. WIND, SURFACE, SPEED/DIRECTION		E		E	E		E	E	E		E	E	E		E	E	D	E	D	D	E	E	E	D	E
71. WIND, SURFACE, GUST SPEED				E			D	E	E		E	E	E		D						E	E			E
72. WIND, SURFACE, GUST SPREAD					E				E		E	E										E			E
73. WIND, UPPER AIR, SPEED/DIRECTION		E			E				E		E	E				E						E			E

Table K-5. Windchill chart.

WIND SPEED (KNOTS)	WIND SPEED (MPH)	LOCAL TEMPERATURE (°F)										
		32	23	14	5	-4	-13	-22	-31	-40	-49	-58
		EQUIVALENT TEMPERATURE (°F)										
	CALM	32	23	14	5	-4	-13	-22	-31	-40	-49	-58
4	5	29	20	10	1	-9	-18	-28	-37	-47	-56	-65
9	10	18	7	-4	-15	-26	-37	-48	-59	-70	-81	-91
13	15	13	-1	-13	-25	-37	-49	-61	-73	-85	-97	-109
17	20	7	-6	-19	-32	-44	-57	-70	-83	-96	-109	-121
21	25	3	-10	-24	-37	-50	-64	-77	-90	-104	-117	-127
26	30	1	-13	-27	-41	-54	-68	-82	-97	-109	-123	-137
30	35	-1	-15	-29	-43	-57	-71	-85	-99	-113	-127	-142
35	40	-3	-17	-31	-45	-59	-74	-87	-102	-116	-131	-145
39	45	-3	-17	-31	-45	-59	-74	-87	-102	-116	-131	-145
43	50	-4	-18	-33	-47	-62	-76	-91	-105	-120	-134	-148
		LITTLE DANGER FOR PROPERLY CLOTHED PERSONS		**CONSIDERABLE DANGER**			**VERY GREAT DANGER**					
DANGER FROM FREEZING OF EXPOSED FLESH												

HOT WEATHER ENVIRONMENT:

In hot weather, important factors are temperature and relative humidity. Primary concerns are physical exhaustion and dehydration. You must consider both because a common work task will take longer and additional water may have to be transported during hot weather. The following are specific concerns:

Water Adequacy. Water supplies and the enforcement of its intake before, during, and after physical activity is critical for survival.

Workload Correlation. Adapt workload or training activities to environmental heat stress conditions.

Rest Periods. Provide adequate break time for physically active personnel in a hot environment.

Chilled Drinking Water. Soldiers will often reject warm drinking water even when they are significantly dehydrated.

Physical Activity. Soldiers' introduction to physical exertion in a hot climate upon arriving from a temperate climate should be as gradual as circumstances allow. They will need 1 to 2 weeks' time to physiologically adjust to the new climate. Lighten physical activity during this period to ensure optimum performance.

Salt. Sodium intake must be maintained in hot weather. Two good meals a day normally provide enough salt for most soldiers.

Humidity. A low WBGT index (e.g., in the morning) may not be a totally safe indicator if the humidity is high. High humidity retards cooling by evaporation of sweat and decreases the urge to drink sufficient water.

Water Spray. A water mist will cool a person. But it should never be substituted for adequate consumption of water before, during, and after strenuous activities in the heat.

The SWO can provide temperature measurement, forecast, and relative humidity forecasts from the following tables. If you need air temperature values measured at your particular location, use the thermometer included in the FALOP BWK.

HOT WEATHER WATER REQUIREMENTS:

Table K-6 shows the water needs for soldiers at 3 activity levels over an 8-hour work period. To determine soldiers' average water needs, you have to know the air temperature and decide the level of activity the troops will be doing. Example: If a soldier is doing 8 hours of hard work in the sun (curve C) when the average temperature for the day is 100°F, his water requirements for the day will be around 15 quarts. This amount of water can be converted into extra weight the soldier must carry. One quart is equal to 2 pounds, so the 15 quarts of water would weigh 30 pounds.

Table K-6. Daily water consumption requirements for three levels of activity.

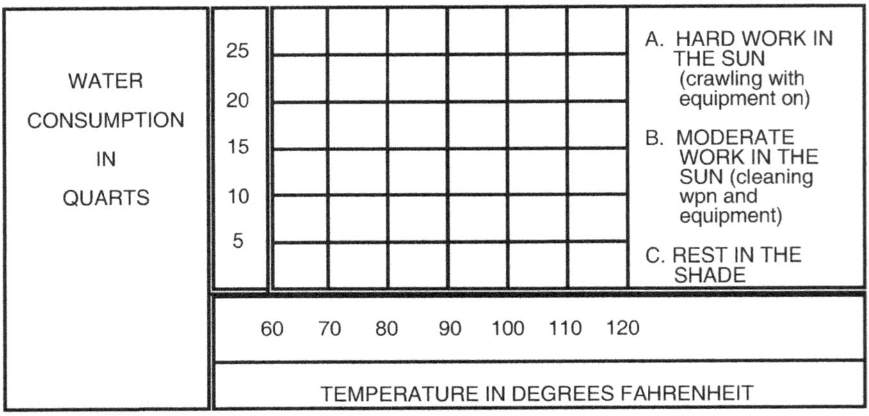

HOT WEATHER WORK TIME LIMITS:

Table K-7 shows the time limits during which work can be performed safely in hot weather. In computing the time limits, consider both air temperature and relative humidity. The values are based on a sitting soldier in BDUs doing light work. Do not consider these tables as absolute limits—especially since light, moderate, or heavy work are difficult to quantify. Table K-8 indicates work time in hot weather based on the WBGT index. Table K-9 shows conversion factors.

Table K-7. Work time in hot weather operations.

TEMPERATURE		RELATIVE HUMIDITY (%)					
(°C)	(°F)	10	30	50	70	90	100
60	140	1 HR	.25 HR				
54	130	2 HR	.5 HR	.25 HR	NO WORK RECOMMENDED		
49	120	4 HR	2 HR	.5 HR	.25 HR		
43	110	12 HR	4 HR	2 HR	.5 HR	.25 HR	
38	100	NO LIMIT	12 HR	4 HR	2 HR	1 HR	.5 HR
32	90	WITH PRECAUTIONS			12 HR	6 HR	4 HR

Table K-8. WBGT index (light work).

HEAT CONDITION/ CATEGORY *	WBGT(°F)	WATER INTAKE QUARTS/HOUR	WORK/REST CYCLE (MINUTES)
1	78.0 to 81.9	AT LEAST ½	CONTINUOUS
2	82.0 to 84.9	AT LEAST 1.2	50/10
3	85.0 to 87.9	AT LEAST 1	45/15
4	88.0 to 89.9	AT LEAST 1-½	30/30
5 **	90.0 and ABOVE	MORE THAN 2	20/40

* MOPP gear or body armor adds at least 10 °F to the WBGT index.
** Suspend PT and strenuous activity. If operational (non-training) mission requires strenuous activity, enforce water intake to minimize expected heat injuries.

Table K-9. Conversion factors.

TO CONVERT:	TO:	USE:
	TEMPERATURE	
Degrees F	Degrees C	^{0}C = 5/9 ^{0}F - 32)
Degrees C	Degrees F	^{0}F = (9/5 ^{0}C) + 32
	DISTANCE	
Kilometers	Miles	0.62
Kilometers	Nautical Miles	0.54
Kilometers	Feet	3,280.80
Miles	Kilometers	1.61
Miles	Nautical Miles	0.87
Miles	Yards	1,760.00
Miles	Feet	5,280.00
Nautical Miles	Kilometers	1.85
Nautical Miles	Miles	1.15
Meters	Feet	3.28
Yards	Feet	3.00
	SPEED	
Kilometers/Hour	Miles/Hour	0.62
Kilometers/Hour	Knots (Nautical Miles/Hour)	0.54
Miles/Hour	Kilometers/Hour	1.61
Miles/Hour	Knots	0.87
Miles/Hour	Feet/Second	1.467
Knots	Kilometers/Hour	1.85
Knots	Miles/Hour	1.15
Meters/Second	Feet/Second	3.281
Meters/Second	Miles/Hour	2.237
	PRESSURE	
Inches of Mercury (Hg)	Millibar	33.86395
Millibar (Mb)	Inches of Mercury	0.295299
	LENGTH	
Feet Meters	0.3048	
Feet Centimeters	30.48	
Inches	Meters	0.0254
Inches	Centimeters	2.54
Meters	Yards	1.094
Yards	Meters	0.9144

GLOSSARY

Section I. Acronyms and Abbreviations

A

A	as acquired (effect used in AGM)
AA	avenue of approach
AAA	air avenue of approach
AAM	air-to-air missile
AAO	Army Aviation Officer
AAR	after-action report
AASLT	air assault
ABCS	Army Battle Command System
abn	airborne
AC	Active Component
ACE	analysis and control element
ACofS	Assistant Chief of Staff
ACR	armored cavalry regiment
ACT	analysis and control team
AD	air defense
ADA	air defense artillery
ADO	air defense officer
ADP	automated data processing
AEB	aerial exploitation battalion
AEPDS	Advanced Electronic Processing and Dissemination System
AF	audio frequency
AFATDS	Advanced Field Artillery Tactical Data System
AFSO	aerial fire support observer
AGL	above ground level
AG/LT	assault gun/light tank
AGM	attack guidance matrix
AI	area of interest
ALO	air liaison officer
ALOC	administrative and logistics center
AMSS	Automatic Meteorological Sensor System
AO	area of operations
AOB	air order of battle
APC	armored personnel carrier
AQF	Advanced QUICKFIX

AR	Army Regulation
ARF	airborne relay facilities
ARFOR	Army force
ARL	airborne reconnaissance low
ARSOF	Army Special Operations Forces
arty	artillery
ASAR	Advanced Synthetic Aperture Radar
ASAS	All-Source Analysis System
ASM	air space management
ASPS	all-source production section
asst	assistant
AT	antitank
ATCCS	Army Tactical Command and Control System
ATGM	antitank guided missile
ATK	attack
ATR	automated target recognition
ATW	antitank weapon
AUTL	Army Universal Task List
auto	automated
avn	aviation
AVTR	airborne video tape recorder
AWACS	Airborne Warning and Control System

B

B	bomber
BAT-D	battlefield deception
BCTP	battle command training program
BDA	battle damage assessment
BDE	brigade
BDU	battle dress uniform
BHL	battle handover line
BIC	battle-induced contaminants
BICC	battalion information control center
BMCT	beginning of morning civil twilight
BMNT	beginning of morning nautical twilight
BMO	battalion maintenance officer
bn	battalion
BOS	battlefield operating system
BP	battle position
BWK	belt weather kit

C

C^2	command and control
C^2I	C^2 and intelligence
C^2W	command and control warfare
C^3	command, control, and communications
C^3I	C^3 and intelligence
C^4I	C^3, computers, and intelligence
C^4I^2	C^4I and information
CA	Civil Affairs/combat assessment
CALL	Center for Army Lessons Learned
CARVER	criticality, accessibility, recuperability, vulnerability, effect, and recognizability
CAS	close air support
Cas Evac	casualty evacuation
CATK	counterattack
CB	counterbattery
cbt	combat
CBU	cluster bomb unit
CCIR	commander's critical information requirements
CCM	cross-country movement
CCMEC	Combined Captured Materiel Exploitation Center
CD	counter-drug
CDMP	combat decision-making
CDR	commander
CEE	captured enemy equipment
CEM	captured enemy materiel
CFE	Conventional Forces In Europe
CFL	coordinated fire line
CFSO	Counterintelligence Force Protection Source Operations
CGS	J-STARS common ground station
CHAALS	Communication High Accuracy Airborne Location System
CHARIOT	Portable S-Band Receive/Transmit Terminal
CHATS	CI/HUMINT Automation Tool Set
chem	chemical
CI	counterintelligence
CIA	Central Intelligence Agency
CINC	commander in chief
CIP	Command Inspection Program
CIR	critical intelligence requirements
CM	collection management
cmd	command

CMEC	Captured Materiel Exploitation Center
CMISE	Corps Military Intelligence Support Element
CMO	civil-military operations
CNR	combat net radio
co	company
COA	course of action
COIN	counterinsurgency
coll	collection
COLT	combat observation lasing team
COMINT	communications intelligence
comm	communications
COMMEX	communications exercise
COMSEC	communications security
CONPLAN	contingency plan
CONUS	continental United States
coord	coordination
COP	command observation post
COR	chief of reconnaissance
COS	critical occupational speciality
CP	command post
CS	combat support
CSBM	Confidence and Security Building Measures
CSP	communications system processor
CSS	combat service support
CSR	controlled supply rate
CTC	combat training center
CTT	commander's tactical terminal
CUCV	commercial utility cargo vehicle

D

D	destroy (effect used in AGM); desirable (weather designator)
DA	Department of the Army
DDMP	deliberate decision-making process
DEA	Drug Enforcement Administration
DF	direction finding
DIA	Defense Intelligence Agency
DIN	Defense Intelligence Network
DISE	Deployable Intelligence Support Element
dissem	dissemination
div	division

DMP	decision-making process
DOCEX	document exploitation
DOD	Department of Defense
DOS	Department of State
DP	decision point
DS	direct support
DSN	Defense Secure Network
DSSCS	Defense Special Security Communications System
DST	decision support template
DSVT	digital subscriber voice terminal
DTG	date-time group
DTSS/MSIP	Digital Topographical Support System/Multi-Spectral Image Processor
DTSS/QRMP	Digital Topographical Support System/QUICK-Response Multicolor Printer
DZ	drop zone
	E
E	essential (weather designator)
EA	electronic attack; engagement area (used in AGM)
EAC	echelons above corps
ECB	echelons corps and below
ECOA	enemy course of action
EECT	end of evening civil twilight
EEFI	essential elements of friendly information
EENT	end of evening nautical twilight
EFVS	electronic fighting vehicle system
e.g.	for example
ELINT	electronic intelligence
EM	electromagnetic
ENGR	engineer
E-O	electro-optical
EP	electronic protection
EPB	electronic preparation of the battlefield
EPDS	Electronic Processing and Dissemination System
EPW	enemy prisoner of war
ERFB-BB	extended-range full-bore-based bleed
ES	electronic warfare support
est	estimated
etc	and so forth
ETRAC	Enhanced Tactical Radar Correlator

ETUT	Enhanced Tactical Users Terminal
EW	electronic warfare
EWO	electronic warfare officer
EWS	early warning system
EWTL	electronic warfare target list
EXEVAL	external evaluation

F

°F	degree Fahrenheit
F	fighter (sortie category)
FA	field artillery
FAC	forward air controller
FAIO	field artillery intelligence officer
FALOP	forward area limited observation program
FARP	forward arming and refueling point
FAST	Forward Area Support Terminal
FAST-I	Forward Area Support Terminal-Improved
FAX	facsimile
FBIS	Foreign Broadcast Information Service
FCE	fire control element
FDA	functional damage assessment
FDC	fire direction center
FDO	fire direction officer
FEBA	forward edge of the battle area
FFIR	friendly forces information requirements
FISTV	first support team vehicle
FLCS	Force Level Control System
FLIR	forward-looking infrared
FLOT	forward line of own troops
flt	flight
FM	field manual
FMC	fully mission capable
FMIB	Foreign Materiel Intelligence Battalion
FO	forward observer
FofF	field of fire
FRAGO	fragmentary order
FS	fire support
FSB	forward support battalion
FSC	fire support center
FSCL	fire support coordination line
FSCM	fire support coordinating measure

FSCOORD	fire support coordinator
FSE	fire support element
FSO	fire support officer
FSOP	field standing operating procedure
FSU	fire support unit
ft	feet
FTI	fixed target indicator
F/W	fixed wing
fwd	forward

G

G2	Assistant Chief of Staff, G2 (Intelligence)
GBCS-(H/L)	ground-based common-sensor (heavy/light)
GCI	ground control intercept
GE	Germany
GHz	gigahertz
GL	grenade launcher
GPS	global positioning system
GRCS	GUARDRAIL Common Sensor
GS	general support
GSM	J-STARS ground station module
GSR	ground surveillance radar

H

H	harrassing fires
HA	humanitarian assistance
HE	high explosive
HF	high frequency
Hg	mercury
HHD	headquarters and headquarters detachment
HHOC	headquarters, headquarters and operations company
HMMWV	high mobility multipurpose wheeled vehicle
HN	host nation
HPT	high-payoff target
HPTL	high payoff target list
HQ	headquarters
hr	hour
HUMINT	human intelligence
HVT	high-value target
hvy	heavy

H/W	hardware
	I
I	immediate (effect used in AGM)
IAW	in accordance with
I&S	intelligence and security
I&W	indications and warnings
ID	identify
IDL	interoperable data link
IEW	intelligence and electronic warfare
IEWCS	Intelligence Electronic Warfare Common Sensor
IEWMAA	IEW mission area analysis
IFV	infantry fighting vehicle
IHL	intelligence handover line
IMA	individual mobilization augmentee
IMETS	Integrated Meteorological System
IMINT	imagery intelligence
incorp	incorporate
INF	Intermediate-range Nuclear Forces intelligence
infil	infiltration
info	information
in/h	inches per hour
INMARSAT	International Maritime Satellite
intel	intelligence
intg	interrogation
INTSUM	intelligence summary
IO	information operations
IPB	intelligence preparation of the battlefield
IPDS	Imagery Processing and Dissemination System
IPF	integrated processing facility
IPS	intelligence production section
IPW	interrogation, prisoners of war
IR	information requirements
I-REMBASS	Improved Remotely Monitored Battlefield Sensor System
ISA	International Standardization Agreement
ISB	intelligence support base
ISM	intelligence synchronization matrix
ISSO	information systems security officer
IVL	intervisibility lines
	J

J2	Intelligence Directorate
JCMEC	Joint Captured Materiel Exploitation Center
JCS	Joint Chiefs of Staff
JIC	Joint Intelligence Center
J-SIIDS	Joint-Service Interior Intrusion Detector System
JSTARS	Joint Surveillance Target Attack Radar System
JTAGS	Joint Tactical Ground Station
JTF	joint task force
JTT/H	joint tactical terminal/hybrid
JTT/H-R3	joint tactical terminal/hybrid-receive only
	K
km	kilometer
	L
LAN	local area network
LAV	light armored vehicle
LC	line of contact
LD	line of departure
LEA	Law Enforcement Agency
LLSO	low-level source operations
LLVI	low-level voice intercept
LMRDFS	Lightweight Man-Transportable Radio Direction Finding System
LNO	liaison officer
LOA	limit of advance
LOB	line of bearing
LOC	line of communication
LOS	line of sight
LP	listening post
LPI	low probability of intercept
LRP	long-range patrol
LRRP	long-range reconnaissance patrol
LRS	long-range surveillance
LRSD	long-range surveillance detachment
LRSU	long-range surveillance unit
LRU	long-range unit
LT	light
LTIOV	latest time information is of value
LZ	landing zone
	M

m	meter
MASINT	measurement and signals intelligence
Mb	millibar
MC	movement corridor
MC&G	mapping, charting, and geodesy
MCOO	modified combined obstacles overlay
MCSF	Mobile Cryptologic Support Facility
MDCI	multidiscipline counterintelligence
MDMP	military decision-making process
MEA	munitions effects assessment
med	medium
MEDEVAC	medical evacuation
MET	meteorological
METL	mission essential task list
METT-T	mission, enemy, terrain and weather, troops, and time available
METT-TC	mission, enemy, terrain and weather, troops, and time available civilians
MEWSS	Mobile Electronic Warfare Support System
MGS	magnetic grid system
mgt	management
MI	military intelligence
MIES	Modernized Imagery Exploitation System
MIJI	meaconing, intrusion, jamming, and interference
min	minute
MIPB	Military Intelligence Professional Bulletin
MIRS	Miniaturization Imagery Receive System
MIST	military intelligence support team
MITT	Mobile Integrated Tactical Terminal
MLRS	multiple launch rocket system
MOPP	mission-oriented protection posture
MP	military police
MR	moon rise
MS	moon set
MSE	mobile subscriber element
MT	medium tank
MTI	moving target indicator
MTOE	modified table of organization and equipment
MTP	mission training plan

N

N	north; neutralize (effect used in AGM)
NA	not applicable
NAI	named area of interest
NATO	North Atlantic Treaty Organization
NBC	nuclear, biological, and chemical
NCA	National Command Authority
NCO	noncommissioned officer
NEO	noncombatant evacuation operation
NFA	no fire area
NGIC	National Ground Intelligence Center
NGLO	naval gunfire liaison officer
NGO	non-governmental organization
NIMA	National Imagery Mapping Agency
NIST	National Intelligence Support Team
NLT	no later than
nmi	nautical miles
NOD	night observation device
NOE	nap of the earth
noncomm	noncommunications
NRT	near-real time
NVD	night vision devices
NVG	night vision goggles
	O
OB	order of battle
OCOKA	obstacles & fields of fire, concealment & cover, key terrain, and avenues of approach
OCONUS	outside continental United States
O/O	on order
OOTW	operations other than war (see SASO)
OP	observation post
OPCON	operational control
OPLAN	operations plan
opns	operations
OPORD	operations order
OPSEC	operations security
OR	operational readiness
OT	operator terminal
	P
P	planned
PAO	public affairs office

PDA	physical damage assessment
PEO	peace enforcement operations
photo	photography
PGM	precision-guided munitions
PIR	priority intelligence requirements
PKO	peacekeeping operations
plt	platoon
PMCS	preventive maintenance checks and services
POL	petroleum, oils, lubricants
PP	passage point
prep	preparation
PSA	post-strike assessment
PSYOP	psychological operations
PT	physical training
PVO	private volunteer organization
PZ	pickup zone
	Q
QSTAG	Quadripartite Standardization Agreement
qty	quantity
	R
R	reconnaissance (sortie category)
R&D	research and development
R&S	reconnaissance and surveillance
RAS	rear area security
RBP	red/black processor
RC	Reserve Components
RCOA	reconnaissance course of action
rcv	receive
RDC	rear detachment commander
recon	reconnaissance
REDTRAIN	readiness training
regt	regiment
REMBASS	Remotely Monitored Battlefield Sensor System
RFI	request for information
RII	request for intelligence information
RISTA	reconnaissance, intelligence, surveillance, and target acquisition
RL	rocket launcher
RNP	radio navigational points (USAF)
ROE	rules of engagement

ROTERM	receive only terminal
RP	release point
rqmts	requirements
RRS	radio relay system
RRT	radio receiver transmission
RRTS	radio relay test set
RSOP	readiness standing operating procedure
RSR	required supply rate
RTO	radiotelephone operator
R/W	rotary wing
RWS	remote workstation

S

S	South; suppress (effect used in AGM)
SA	small arms
S&TI	Scientific & Technical Intelligence
S1	Adjutant (US Army)
S2	Intelligence Officer (US Army)
S3	Operations and Training Officer (US Army)
S4	Supply Officer (US Army)
S5	Civil Affairs Officer (US Army)
SAEDA	Subversion and Espionage Directed Against the US Army
SALUTE	size, activity, location, unit, time, and equipment
SAM	surface-to-air missile
SAR	synthetic aperture radar
SASO	stability and support operations (formerly OOTW)
SATCOM	satellite communications
SCI	sensitive compartmented information
SDO	staff duty officer
SDS	security data system
SEAD	suppression of enemy air defenses
sep bde	separate brigade
SF	Special Forces
SI	special intelligence
SID	secondary imagery dissemination
SIDS	Secondary Imagery Dissemination System
SIGINT	signals intelligence
SIGINT/EW	signals intelligence/electronic warfare
SIGO	signal officer
SII	statement of intelligence interest

SINCGARS	Single-Channel Ground and Airborne Radio System
SIR	specific information requirements
SITMAP	situation map
SJA	Staff Judge Advocate
SLAR	side-looking airborne radar
SOCCE	Special Operations Command and Control Element
SOF	special operations forces
SOFA	Status of Forces Agreement
SOI	signals operating instructions
SOP	standing operating procedure
SOR	specific orders and requests
SOT-A	support operations team-Alpha
SOT-B	support operations team-Bravo
SP	start point
SPIRIT	Special Purpose Integrated Remote Intelligence Terminal
sqdn	squadron
SR	sunrise
SS	sunset
SSM	surface-to-surface missile
SSO	special security office
SSP	single-soure processing
ST	student text
STANAG	Standardization Agreement
STU III	Secure Telephone Unit-Third Generation
SUCCESS	Synthesized UHF Computer Controller Enhanced Subsystem
survl	surveillance
S/W	software
SWO	staff weather officer

T

TA	target acquisition
TAA	tactical assembly area
TAB	target acquisition battery
tac	tactical
TAC	tactical command post
TACFIRE	tactical fire direction system
TACLINK	tactical link
TACSAT	tactical satellite

TADIXS-B	Tactical Data Information Exchange System-Broadcast
TAI	target area of interest
TAREX	target exploitation
TB	tank brigade
TD	tank division
TDA	target development assessment
TDDS	TRAP Data Dissemination System
TDMP	tactical decision-making process
TDOA	time differential of arrival
TECHINT	technical intelligence
TENCAP	Tactical Exploitation of National Capabilities
TF	task force
tgt	target
THAADS	Theater High Altitude Area Defense System
THMT	tactical high mobility terminal
TIBS	Tactical Information Broadcast System
TLE	target location error
TLP	troop-leading procedures
TOC	tactical operations center
TOPS	Tactical Onboard Processing System (for TADIXS-B)
TOW	tube-launched, optically tracked, wire-guided
TPL	time phase line
TRAP	Tactical Related Applications
TRIXS	Tactical Reconnaissance Intelligence Exchange System
trng	training
TROJAN SPIRIT	Special Purpose Integrated Remote Intelligence Terminal
TRRIP	Theater Rapid Response Intelligence Package
TSA	target system assessment
TSOP	tactical standing operating procedure
TSS	target selection standards
TTP	tactics, techniques, and procedures
TUAV	tactical unmanned aerial vehicle
TVA	target value analysis

U

UAV	unmanned aerial vehicle
UBL	unit, basic load
UHF	ulltra high frequency
USAF	US Air Force
USMTF	United States Message Text Formats
USSID	United States Signal Intelligence Directive

UW	unconventional warfare
	V
VHF	very high frequency
vis	visibility
	W
w	with
WARNO	warning order
WATCHCON	watch condition
WBGT	wet-bulb-globe temperature
WETM	weather team
wkstn	workstation
wpn	weapon
	X
XO	executive officer
XPLN	exploitation
xmit	transmit

Section II. Terms.

Global security forecast A forecast of where, when, and between whom conflicts can occur.

Battlefield development plans A 5- to 20-year forecast of how selected threat forces will organize and operate on the battlefield.

Automated and hardcopy databases Databases such as the Stylized Country Force Assessments and the Army Force Planning and Data Assumptions which provide detailed projections on OB, equipment, logistics, and readiness (ground, air, naval forces) of potential threats 2 to 25 years into the future.

TECHINT and User Bulletins. Provide a high resolution view of threat equipment and weapons systems and capabilities, and complement traditional OB information. They can be obtained from NGIC.

Open source studies and articles State Department background papers, FBIS, public libraries, universities, and corporations with contracts in countries of interest). Databases of sources need to be researched and exercised in advance of mission execution.

Other services. The USMC, other nation's information, fact books, and fact sheets on tactics, terrain, and country studies.

REFERENCES

SOURCES USED

These are the sources quoted or paraphrased in this publication.

AR 381-10	*US Army Intelligence Activities* 1 Jul 84.
FM 6-20-10	*Tactics, Techniques, and Procedures for the Targeting Process (MCRP 3-1.6.14)* 8 May 96
FM 17-95	*Cavalry Operations.* 24 Dec 96.
FM 33-1	*Psychological Operations (FMFM 3-53)* 18 Feb 93.
FM 34-1	*Intelligence and Electronic Warfare Operations* 27 Sep 94.
FM 34-2	*Collection Management and Synchronization Planning* 8 Mar 94.
FM 34-2-1	*Tactics, Techniques, and Procedures for Reconnaissance and Surveillance and Intelligence Support to Counterreconnaissance* 19 Jun 91.
FM 34-3	*Intelligence Analysis.* 15 Mar 90.
FM 34-8	*Combat Commander's Handbook on Intelligence.* 28 Sep 92.
FM 34-25-3	*All-Source Analysis System and the Analysis and Control Element.* 3 Oct 95.
FM 34-37	*Echelons Above Corps (EAC) Intelligence & Electronic Warfare (IEW) Operations.* 15 Jan 91.
(U) FM 34-40	*Electronic Warfare Operations,* 9 Oct 87, Secret.
FM 34-130	*Intelligence Preparation of the Battlefield.* 8 Jul 94.
FM 41-10	*Civil Affairs Operations.* 11 Jan 93.
FM 100-6	*Information Operations.* 27 Aug 96.
FM 100-17	*Mobilization, Deployment, Redeployment, Demobilization.* 28 Oct 92.
FM 101-5	*Staff Organizations and Operations.* 25 May 84.
FM 101-5-1	*Operational Terms and Symbols.* 21 Oct 85.
Joint Pub 1-02	*Department of Defense Dictionary of Military and Associated Terms.* 23 Mar 94.
Joint Pub 6-04.11	*US Message Text Formatting Program Repository of USMTF Program Items for US Implementation Guidance.* 1 Oct 91.

DOCUMENTS NEEDED

These documents must be available to the intended users of this publication.

AR 190-11	*Physical Security of Arms, Ammunition and Explosives.* 30 Sep 93.
AR 190-13	*The Army Physical Security Program.* 30 Sep 93.

AR 190-14	*Carrying of Firearms and Use of Force for Law Enforcement and Security Duties.* 12 Mar 93.
AR 190-51	*Security of Unclassified Army Property (Sensitive and Nonsensitive).* 30 Sep 93.
AR 525-13	*The Army Combatting Terrorism Program.* 25 Feb 88.
AR 380-5	*Department of the Army Information Security Program.* 25 Feb 88.
AR 380-10	*Technology Transfer, Disclosure of Information and Contacts with Foreign Representatives.* 30 Dec 94.
AR 380-19	*Information Systems Security.* 1 Aug 90.
AR 380-67	*The Department of the Army Personnel Security Program.* 9 Sep 88.
AR 381-1	*Controls of Dissemination of Intelligence Information.* 12 Feb 90.
AR 381-12	*Subversion and Espionage Directed Against the US Army (SAEDA).* 15 Jan 93.
AR 381-20	*The Army Counterintelligence Program.* 15 Nov 93.
(U) AR 381-172	Counterintelligence Force Protection Source Operations *(CFSO) and Low Level Source Operations (LLSO).* Secret.
AR 530-1	*Operations Security (OPSEC).* 3 Mar 95.
FM 1-111	*Aviation Brigades.* 27 Aug 90.
FM 3-4	*NBC Protection (FMFM 11-9) (reprinted w/basic Incl C1).* 29 May 92.
FM 3-5	*NBC Contamination (FMFM 11-10).* 17 Nov 93.
FM 3-50	*Smoke Operations.* 4 Dec 90.
FM 5-30	*Engineer Intelligence (reprinted w/Basic Incl C1).* 22 Sep 67.
FM 5-34	*Enginer Field Data.* 14 Sep 87.
FM 5-36	*Route Reconnaissance and Classification.* 10 May 85.
FM 6-20	*Fire Support in the AirLand Battle* 17 May 88.
FM 5-100	*Engineer Operations.* 27 Feb 96.
FM 5-101	*Mobility.* 23 Jan 85.
FM 5-102	*Countermobility.* 14 Mar 85.
FM 5-103	*Survivability.* 10 Jan 85..
FM 6-121	*Tactics, Techniques, and Procedures for Field Artillery Target Acquisition.* 25 Sep 90.
FM 7-7	*The Mechanized Infantry Platoon and Squad.* 15 Mar 85.
FM 7-7J	*Mechanized Infantry Platoon & Squad (BRADLEY).* 7 May 93.
FM 7-8	*Infantry Rifle Platoon and Squad.* 22 Apr 92.
FM 7-10	*The Infantry Rifle Company.* 14 Dec 90.
FM 7-20	*The Infantry Battalion.* 6 Apr 92.
FM 7-70	*Light Infantry Platoon & Squad.* 10 Sep 86
FM 7-71	*Light Infantry Company.* 28 Aug 87.
FM 7-92	*The Infantry Reconnaissance Platoon and Squad (Airborne, Air Assault, Light Infantry).* 23 Dec 92.
FM 7-93	*Long-Range Survival Unit Operations* 3 Oct 95.

FM 10-14-2	*Guide for the Battalion S4 (reprinted w/Basic Incl C1.)* 30 Dec 81.
FM 10-14-3	*Surveying Officer's Guide.* 30 Dec 81.
FM 11-30	*MSE Communications in the Corps/Division.* 27 Feb 91.
FM 11-32	*Combat Net Radio Operations.* 15 Oct 90.
FM 11-37	*MSE Primer for Smal-Unit Leaders.* 14 Nov 90.
FM 11-50	*Combat Communications Within the Division (Heavy and Light).* 4 Apr 91.
FM 17-15	*Tank Platoon.* 3 Apr 96.
FM 17-95	*Cavalry Operations.* 24 Dec 96.
FM 17-95-10	*The Armored Cavalry Regiment and Squad.* 22 Sep 93.
FM 17-97	*Cavalry Troop.* 3 Oct 95.
FM 17-98-1	*Scout Leader's Handbook.* 24 Sep 90.
FM 19-1	*Military Police Support for the AirLand Battle.* 23 May 88.
FM 19-30	Physical Security. 1 Mar 79.
FM 19-40	*Enemy Prisoners of War, Civilian Internees and Detained Persons.* 27 Feb 76.
FM 21-26	*Map Reading and Land Navigation.* 7 May 93.
FM 21-31	*Topographical Symbols (reprinted w/Basic Incl C1).* 19 Jun 61.
FM 25-100	*Training the Force.* 15 Nov 88.
FM 25-101	*Battle Focused Training.* 30 Sep 90.
(U) FM 34-5	*Human Intelligence and Related Counterintelligence Operations,* 29 Jul 94, Secret.
FM 34-7	*Intelligence and Electronic Warfare Support to Low-Intensity Conflict Operations.* 18 May 93.
FM 34-10	*Division Intelligence and Electronics Warfare Operations.* 25 Nov 86.
FM 34-13	*Military Intelligence (MI) Battalion, Combat Electronic Warfare Intelligence (CEWI) (Div), Leaders Handbook for Platoons and Teams.* 15 Apr 83.
FM 34-25	*Corps Intelligence and Electronic Warfare Operations.* 30 Sep 87.
FM 34-25-1	*Joint Surveillance Target Attack Radar System (JOINT STARS).* 3 Oct 95.
(U) FM 34-25-6	*Tactics, Techniques, and Procedures for the Electronic Processing & Dissemination System* ,15 Jun 92, Secret.
(U) FM 34-25-7	Special Electronic Mission Aircraft (SEMA) Survivability 3 Oct 95, Secret.
FM 34-35	*Armored Cavalry Regiment (ACR) and Separate Brigade Intelligence and Electronic Warfare (IEW) Operations.* 12 Dec 90.
FM 34-36	*Special Operation Forces IEW Operations* 30 Sep 91.
(U) FM 34-40-3	*Tactical Signals Intelligence (SIGINT) Analysis Operations,* 21 May 91, Confidential

FM 34-40-7	*Communications Jamming Handbook.* 23 Nov 92.
FM 34-40-9	*Direction Finding Operations.* 29 Aug 91.
(U) FM 34-40-12	*Morse Code Intercept Operations* ,26 Aug 91, Confidential.
(U) FM 34-40-13	*Electronic Intelligence (ELINT) Operations*,10 Sep 91, Secret.
FM 34-52	*Intelligence Interrogation.* 28 Sep 92.
FM 34-54	*Battlefield Technical Intelligence.* 5 Apr 90.
FM 34-60	*Counterintelligence.* 3 Oct 95.
FM 34-80	*Brigade and Battalion Intelligence and Electronic Warfare Operations.* 15 Apr 86.
FM 34-81	*Weather Support for Army Tactical Operations (AFM 105-4).* 31 Aug 89
FM 34-81-1	*Battlefield Weather Effects.* 23 Dec 92.
FM 34-82	*Military Intelligence Unit Exercise Development Guide.* 26 Nov 85.
FM 71-3	*Armored and Mechanized Infantry Brigade.* 8 Jan 96.
FM 71-100	*Division Operations.* 28 Aug 96.
FM 90-2	*Battlefield Deception.* 3 Oct 88.
(U) FM 90-2A	*Electronic Deception*,12 Jun 89, Secret.
FM 90-7	*Combined Arms Obstacle Integration.* 29 Sep 94.
FM 90-14	*Rear Battle.* 10 Jun 85.
FM 90-29	*Noncombatant Evacuation Operations.* 17 Oct 94.
FM 100-2-2	*The Soviet Army: Specialized Warfare and Rear Area Support.* 16 Jul 84.
FM 100-2-3	*The Soviet Army: Troops, Organization and Equipment.* 16 Jun 91.
FM 100-5	*Operations.* 14 Jun 93.
FM 100-13	*Battefield Coordination Detachment (BCD).* 5 Sep 96.
FM 100-19	*Domestic Support Operations. (FMFM 7-10)* 1 Jul 93.
FM 100-20	*Military Operations in Low-Intensity Conflict (AFP 3-20).* 5 Dec 90.
FM 100-23	*Peace Operations.* 30 Dec 94.
FM 100-25	*Doctrine for Army Special Operations Forces.* 12 Dec 91.
FM 101-10-1/1	*Staff Officer's Field Manual-Organizational, Technical, and Logistical Data (Volume 1).* 7 Oct 87.
FM 101-10-1/2	*Staff Officer's Field Manual-Organizational, Technical, and Logistical Planning Factors (Volume 2) (reprinted w/Basic Incl C1).* 7 Oct 87.
ST 101-5	*Command and Staff Decision Processes*
ST 101-9	*The Tactical Decisionmaking Process*
TC 12-17	*Adjutant's Call/The S1 Handbook.* 17 Mar 92.
TC 34-55	*Imagery Intelligence.* 3 Oct 88.
Joint Pub 1	*Joint Warfare for the U.S. Armed Forces.* 10 Jan 95.
Joint Pub 2-0	*Joint Doctrine for Intelligence Support to Operations.*

	5 May 95.
Joint Pub 3-0	*Doctrine for Joint Operations.* 1 Feb 95.
Joint Pub 5-0	*Doctrine for Planning Joint Operations.* 13 Apr 95.
Joint Pub 6-02	*TAC C4 Systems*
(U) Joint Pub 6-02	*Joint Doctrine for Employment of Operational/Tactical Comand, Control, Communications, and Computer System,* 1 Oct 96, Secret.
DA Form 2028	*Recommended Changes to Publications and Blank Forms.* 1 Feb 74.
DA Form 3964	*Classified Document Accountability Record.* Jul 79.

Made in the USA
Monee, IL
04 November 2026

67383142R00118